Electromagnetic Waves 2

SCIENCES

Waves, Field Directors – Pierre-Noël Favennec, Frédérique de Fornel

Electromagnetism, Subject Head – Pierre-Noël Favennec

Electromagnetic Waves 2

Antennas

Coordinated by

Pierre-Noël Favennec

iSTE

WILEY

First published 2020 in Great Britain and the United States by ISTE Ltd and John Wiley & Sons, Inc.

ISTE Ltd
27-37 St George's Road
London SW19 4EU
UK

www.iste.co.uk

John Wiley & Sons, Inc.
111 River Street
Hoboken, NJ 07030
USA

www.wiley.com

Library of Congress Control Number: 2020937434

British Library Cataloguing-in-Publication Data
A CIP record for this book is available from the British Library
ISBN 978-1-78945-007-1

ERC code:
PE2 Fundamental Constituents of Matter
PE2_6 Electromagnetism

Contents

Preface

Pierre-Noël Favennec
ArmorScience, Lannion, France

Any electric charge set in motion produces electromagnetic radiation which propagates in space. This property is the basis of radioelectric, or photonic radiation production, used in particular in radio, television and communication systems among others. Any system supplied with electricity, or any element provided with electric charge, emits electromagnetic radiation and generates an electric and/or magnetic field in its close, or even distant, vicinity which is known as an "electromagnetic field".

Before Maxwell's work, we understood physical reality in terms of material points. After it, we represented physical reality with continuous fields. The concept of a field finds its origin, and its name, in the idea of describing a physical phenomenon from an underlying medium, which would explain the physical properties of space (a field of forces for a field of wheat subjected to the wind). Following Maxwell's research, the fields acquired an autonomous existence and reached the status of physical beings in their own right, no longer describing "the place where" but "the thing that". This movement was largely supported by the development of the mathematical formalism of the fields, in terms of partial differential equations. This, with regard to electricity and magnetism, is the content of Maxwell's theory which he published in 1861.

Maxwell is one of the greatest scientists, who changed our view of the world. He made a decisive contribution to the unifying and synthetic vision of electricity and

For a color version of all figures in this book, see www.iste.co.uk/favennec/electromagnetic2.zip.

Electromagnetic Waves 2,
coordinated by Pierre-Noël Favennec.

magnetism. He showed that two domains, that of electric charges and their interactions and that of currents and magnetism, were only two facets of the same problem. Synthesized by four equations combining in the same formalism their respective characteristic magnitudes. He stated these interactions in clear mathematical language: Maxwell's equations. The vision of a universe formed by particles was succeeded by a world governed by fields, acting from a distance.

The electromagnetic field is the set of vector fields $(\vec{E}, \vec{B})$. The properties of the electromagnetic field at a point in space are determined by the properties of the electric field $\vec{E}$ and the magnetic field $\vec{B}$ at a point. In physics, the term "field" refers to the situation where we are in the presence of a physical magnitude distributed in a given region of space. This magnitude has a value determined at each point in this space and at all times. Having an area of space where there is an electromagnetic field, means that at each point in this space, we have two vector variables $\vec{E}$ and $\vec{B}$.

Electromagnetic waves are produced by excited matter. The deexcitation of the excited source produces around it a periodic variation of the electromagnetic field which propagates gradually in the vacuum at the phase speed (or propagation speed) close to 300,000 km per second. Depending on their emission frequency domain, they have different names: radio waves for the lowest frequencies, infrared waves, visible optical waves, then ultraviolet, then for the highest frequencies, X-rays and gamma rays. The electromagnetic wave propagates: a variable electric field generates a variable magnetic field and conversely a variable magnetic field generates a variable electric field. The conjoint propagation of these variations in a region constitutes a continuous wave phenomenon, capable of propagating (across the vacuum at 300,000 kilometers per second), transporting energy without the need for material support.

Waves are vibrations that propagate from one place to another in space, in a material medium or in a vacuum. Electromagnetic vibrations (electromagnetic waves) are waves obeying the laws of electromagnetism. Mechanical vibrations (pendulum, acoustics, etc.) obey the laws of mechanics, but often these mechanical vibrations are in fact fundamentally electromagnetic, due to the electromagnetic interactions of atoms and molecules of materials; they are described by "approximate" laws according to movements following the laws of mechanics.

Characterizing or measuring an electromagnetic field is carried out via current or voltage measurements. The electromagnetic field located in one place is the set of vector fields ($\vec{E}$, $\vec{B}$). *Sensors or antennas* measure, at a given point, the currents or voltages resulting from the field from the different vector magnitudes $\vec{E}$ and $\vec{B}$. Subsequent processing can, if useful, select the different frequencies.

In our everyday life, the *environmental electromagnetic field* does not arise from a single source. There are fields of natural origin (the sun, galaxy, geomagnetism, etc.) and those of human origin (household materials, transport, telecommunications, energy supply, etc.). Each point on the planet is subjected to a fairly intense electromagnetic "bath" depending on its location. The drawing below, envisaged by Michel Urien, shows that we are all "willingly" bathed in these electromagnetic waves. Let us try to understand our *electromagnetic environment*!

Figure P.1. *A wave bath envisaged by Michel Urien*[1]

This referenced work, presented in two inseparable volumes, is essential for any student, engineer or researcher wishing to understand electromagnetism and all the technologies derived from it.

Volume 1 is oriented towards the basic phenomena explaining electromagnetism: the famous Maxwell equations – essential to know – then the propagation phenomena of electromagnetic waves. It only concerns non-ionizing radiation, which is radiation from waves whose energies are insufficient to ionize an atom, that is to say incapable of removing an electron from matter. This excludes all radiation with an energy greater than 12.4 eV, that is that generated by X-ray and gamma ray emitters. This work is made up of two chapters.

1 Source: www.armorscience.com.

In **Chapter 1, Ibrahima Sakho** presents the Maxwell equations as clearly as possible. These equations are essential to comprehensively approach electromagnetism and all its derived fields such as radioelectricity, photonics, geolocation, measurement, telecommunications, medical imagery, radio astronomy, etc.

In **Chapter 2, Hervé Sizun** describes the propagation phenomena of electromagnetic, radio and photonic waves. Many factors, often complex, must be taken into account to properly understand these propagation problems in free and sometimes confined spaces.

In **Volume 2, Jean-Pierre Blot**, expert in *radio antennas* of all configurations, directs his analysis towards antennas, essential elements for the detection of electromagnetic waves, their characterization and use. This volume is intended to describe what an effective antenna should be, according to various parameters and conditions of use. It does not address the detection problems specific to photonics. Photonics and these detection problems will be seen in a future publication of the "Waves" series.

Important appendices with essential information, presenting in particular mathematical tools, complete these two volumes.

References

Cartini, R. (1993). *Panorama encyclopédique des sciences*. Belin, Paris.

de Fornel, F., Favennec, P.-N. (eds) (2007). Mesures en électromagnétisme. *Revue RS série I2M*, 7(1–4).

Favennec, P.-N. (2008). *Mesures de l'exposition humaine aux champs radio-électriques – Environnement radioélectrique*. Techniques de l'Ingénieur, Saint-Denis.

Serres, M., Farouki, N. (1997). *Dictionnaire des sciences*. Flammarion, Paris.

1

General Information on Antennas

Jean-Pierre BLOT
EUROSATCOM, Les Ulis, France

1.1. Definition, context, and regulation[1]

In radioelectricity, an *antenna* (using the IEEE Standard 145-1993 definition[2]) is a device which makes it possible to radiate (transmitter) or to collect (receiver) electromagnetic waves in an efficient way. Heinrich Hertz (1857–1894) used antennas for the first time in 1889, to demonstrate the existence of electromagnetic waves, predicted by James Clerk Maxwell's theory. He used doublet antennas for both reception and transmission. He even installed a dipole transmitter at the focal point of a parabolic reflector. The work and drawings of the invention were

For a color version of all figures in this book, see www.iste.co.uk/favennec/electromagnetic2. zip.

1 Inspired by the *Lettre de l'Autorité* from the *Autorité de Régulation des Communications Électroniques et des Postes*, no. 46, September/October 2005, and ITU-R reports.

2 The IEEE standard documents are developed by the Technical Committees of the IEEE societies and the Standards Coordinating Committees, from the office of the IEEE Standards Board. Committee members are volunteers without compensation, and are not necessarily members of the institute. The standards developed within the institute represent a consensus of experts on a subject that expresses a particular interest in the development of a standard. The use of an IEEE standard is optional. Its existence, as with the ITU, does not exclude other ways of making, testing, measuring, and placing an item on the market. IEEE standards are subject to review or reaffirmation every five years. It is therefore important to make sure you have the latest version.

Electromagnetic Waves 2,
coordinated by Pierre-Noël FAVENNEC.

published in the *Annalen der Physik und Chemie*[3]. The term "antenna" was first used by Guillermo Marconi (1874–1937).

Today, the technological evolution of antennas is inseparable from that of satellites. It was during the Second World War that intensive development was undertaken in this field for radar, and is behind many technologies used today. It was with the invention of triode tube generators, in the 1920s, then magnetrons and klystrons, operating at 1 gigahertz for radar, that it all started: antennas with reflectors, lenses, networks of slotted waveguides, wire antennas, etc. It gained a new impetus with the advent of space exploration, which started with the launch of Sputnik in 1957, and which did not stop evolving with the demand for new telecommunications services, which were first fixed, then mobile, with geostationary satellites that were initially passive and then active and low-orbiting.

The evolution of new technologies, such as reconfigurable printed antennas, has made it possible to produce satellites to travel towards high, medium, and low orbits (Galileo, Iridium, Global Star, etc.). These new-generation satellites are light and less expensive and launched and in batches, by launchers which are also lighter. At the same time, demands for new frequency bands are gradually moving towards higher and higher frequencies.

The possibilities offered by the propagation of electromagnetic waves (EM) in the natural environment are utilized for multiple purposes, including civil and military purposes: broadcasting, television, radar, telecommunications, radio navigation, etc. In all these applications, the antenna is the essential component for radiating and capturing waves. They were, therefore, particularly suitable for radiating wire structures, used by the pioneers of radiotelegraphy since the end of the 19th Century. Since then, antenna technologies have evolved with the multiplication of communication needs, and have diversified to take into account both the discovery of new terrestrial propagation modes as well as constant technological progress towards increasingly higher frequencies.

The antenna designation prevails over a large part of the electromagnetic wave spectrum, which ranges from wavelengths in kilometers to submillimeter wavelengths which are close to infrared (Table 1.1). At higher wavelengths, in the field of optics, the universal term of "antenna" is not used, allowing for a more precise denomination of the various constituent elements (lenses, for example). The field of antennas concerned here is limited technologically, and in an internationally accepted manner, to that of the classification and main characteristics of the

3 *Annalen der Physik und Chemie* is the oldest scientific journal in physics, published since 1790. Here, we are referring to Volume 36, published in 1889.

electromagnetic waves given in Table 1.1: we will discuss waves and microwaves below infrared, and radiation above this range that is no longer the field of telecommunications. Technologies in the field of optics will be covered in more specific works on photonics.

The two aspects of antenna function, transmiting and receiving, are closely related. It is the electronic equipment, associated with an antenna, that defines its possible specificity. In almost all cases, the same antenna is used for transmission as for reception. This is a consequence of the reciprocity theorem. In some cases, when the antennas contain non-linear materials, they are not reciprocal. Because of the reciprocity, there will hardly ever be a difference between the radiation in emission or reception. The qualities stated for an antenna below will be so in both operating modes, without this being specified.

For its purpose, the radiation of an antenna is reduced, by its illumination, its emitted power, and the sensitivity of its receiver, to a problem of covering the surrounding or geographical space at small, medium, or large distances: this may mean covering the surrounding space evenly, or a very localized area in this space. Such a concern allows the first functional classification, to distinguish antennas with little or no directionality from directional antennas..

Quickly, with the multiplication of applications, be they spatial, terrestrial or wireless, rules for coexistence, coverage and frequency management need to be established. Frequency may physically be the opposite of time ($\nu = 1/t$), it is an economically rare and exhaustible good that cannot be produced. In addition, it is a heterogeneous good; the "low" frequencies (below 1 gigahertz, see Figure 1.1) have a use value that is higher than that of the "high" frequencies, because their property of good propagation allows for less expensive coverage in lightly-populated areas and better penetration into buildings in urban areas. To best manage the scarcity and heterogeneity of the hertzian spectrum, it should be allocated in the most dynamic and flexible way possible.

Firstly, it should be allocated in a dynamic way because the sustained innovation and technological progress that drive the electronic communications sector, as well as the rapid growth of the markets, transform and renew the range of spectrum use. Some technologies are expected to decline and die out in the medium to long term, such as the GSM or analog television, while others emerge or develop in favor of everything digital, such as UMTS, WiMax, or TNT.

International nomenclature	Frequency	Wavelength	Other names	Examples of use
TLF (*Tremendously Low Frequency*)	0 Hz to 3 Hz	100,000 km to ∞	–	Magnetic fields, waves and natural electromagnetic noises, etc.
ELF (*Extremely Low Frequency*)	3 Hz to 30 Hz	100,000 km to 10,000 km	–	Detection of natural phenomena, brain waves, research in geophysics, molecular spectral rays, etc.
SLF (*Super Low Frequency*)	30 Hz to 300 Hz	10,000 km to 1000 km	–	Detection of natural phenomena, communication with submarines, waves from power lines, industrial inductive uses, etc.
ULF (*Ultra Low Frequency*)	300 Hz to 3000 Hz	1000 km to 100 km	–	Detection of natural phenomena, human physiological waves, electrical waves of telephone networks, etc.
VLF (*Very Low Frequency*)	3 kHz to 30 kHz	100 km to 10 km	Myriametric waves	Wireless communications with submarines, medical implants, scientific research, radio telecommunications, etc.
LF (*Low Frequency*)	30 kHz to 300 kHz	10 km to 1 km	Low frequency or long wave or kilometric wave	Amateur radio, radio navigation, longwave broadcasting, radio-frequency identification, etc.
MF (*Medium Frequency*)	300 kHz to 3 MHz	1 km to 100 m	Hectometric or medium wave or frequency	Amateur radio, AM radio, maritime services, air communications, etc.

International nomenclature	Frequency	Wavelength	Other names	Examples of use
HF (***High Frequency***)	3 MHz to 30 MHz	100 m to 10 m	Short-wave or decametric waves	Various organizations, including the military, international broadcasting, maritime, aeronautics, amateur radio, weather, radio emergency communications, CPL transmission, etc.
VHF (***Very High Frequency***)	30 MHz to 300 MHz	10 m to 1 m	Ultra-short waves or metric waves	Television, broadcasting, land-to-air transmission, air-to-air transmission, terrestrial and maritime mobile communication, amateur radio, meteorology, etc.
UHF (***Ultra High Frequency***)	300 MHz to 3 GHz	1 m to 10 cm	Decimetric waves	Private networks, the military, GSM, GPS, Wi-Fi, television, amateur radio, satellite, tropospheric links, etc.
SHF (***Super High Frequency***)	3 GHz to 30 GHz	10 cm to 1 cm	Centimeter waves	Private networks, Wi-Fi, satellite, satellite broadcasting (TV), radio-relay systems, radar, terrestrial and satellite links, etc.
EHF (***Extremely High Frequency***)	30 GHz to 300 GHz	1 cm to 1 mm	Millimeter waves	Private networks, collision avoidance radar for cars, transportable video links, space research, etc.
THz (***terahertz***)	300 GHz to 3 THz	1 mm to 100 µm	Decimillimeter waves	Infrared waves

Table 1.1. *Official ITU nomenclature of frequency bands*

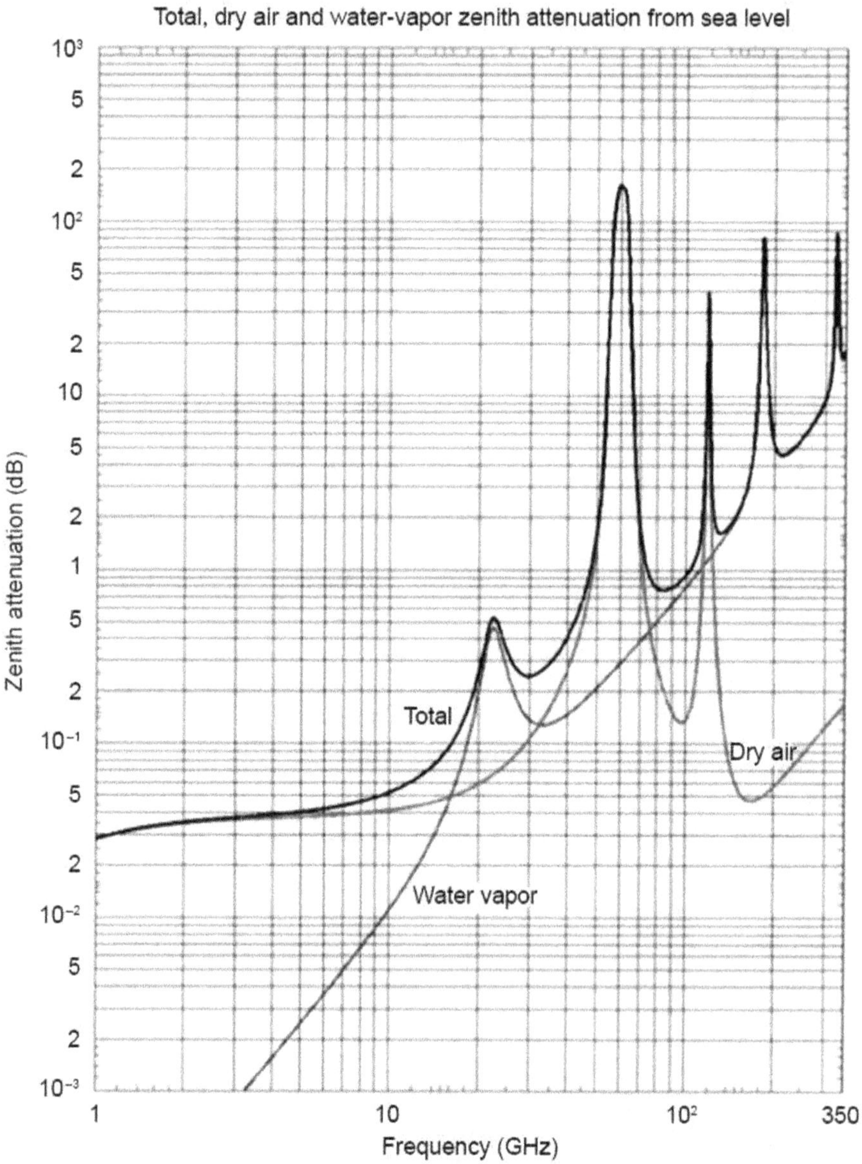

Figure 1.1. *Total zenith attenuation in free space under clear sky (ITU-R report 676)*

Secondly, it should be allocated in a flexible way because the growing convergence between mobile, fixed, audiovisual, and telecommunications services, as well as the use of hybrid systems using, in particular, a satellite segment and a ground segment, would make partitioned allocation ineffective, reserving frequency bands too rigidly to a certain type of service or technical means. A careful trade-off therefore must be made between the concern for neutrality, favoring flexibility, and maintaining a certain degree of harmonization at European, or continental, and global levels, guaranteeing coherence.

The radio spectrum is, indeed, a finite resource, and even if it extends up to a few hundreds of gigahertz, frequencies below a few tens of gigahertz are practically the only ones to be used. Indeed, the laws of physics state that the range of radio communications decreases very quickly with frequency, which discourages the use of high frequencies, and *a fortiori* the development of the corresponding components. Furthermore, the propagation of EM waves in free space is selective, due to the presence of oxygen (O_2) and water vapor, as clouds, which favors the use of frequencies below a few tens of gigahertz (Figure 1.1).

The success of innovation is therefore based on a compromise between the technological and commercial risk linked to the increase in frequency, its additional costs, and the regulatory risk linked to the need to share with existing systems. It is generally this latter risk that is preferred, as evidenced by recent developments (5 GHz Wi-Fi, shared with defense and weather radar, automobile radar at 23 GHz shared with radio-relay systems, police radar with Earth observation satellites).

Remedies for scarcity therefore depend on an appropriate regulatory framework. This framework has to have an international component, because radio waves do not stop at borders. There is, therefore, a need for international coordination. Globally, the RR (Radio Regulations) of the ITU (International Telecommunication Union) allocates frequency bands to "services" defined by broad categories, as summarized in Table 1.1. When several services share the same band, it is necessary to define priorities among users and the technical and regulatory conditions for sharing.

It is in sharing the spectrum that new technologies provide solutions such as "smart" antennas, capable of adapting to their radio environment, which will bring about new mobility technologies, for example, for 4G and 5G. Wireless broadband access systems make it possible to connect to the Internet at high speed by radio waves, but they present the disadvantage for the user that the latter has to install a "fixed" antenna, oriented towards one of the base stations of their service provider. The development of innovative signal modulation techniques, associated with the use of "smart" antennas, opens up prospects of seeing this initially fixed service evolve towards mobile applications, initially on hotspots and then mobiles. Among the new modulation techniques, the OFDM (Orthogonal Frequency-Division Multiplexing) foreshadowed what "4G" is. The data flow, which represents a high bit rate, is distributed over a series of modulated subcarriers at low bit rates. A more efficient technique, it facilitates broadband reception without requiring the installation of a fixed antenna in direct visibility of the operator's base station. At the base stations, smart antennas are capable of dynamically splitting the transmitted signal into several narrow beams directed towards the users and at mobiles, thereby increasing the carrier-to-noise ratio (C/N). This improvement offers multiple advantages, both in terms of the use of spectral resources and in increasing the range of cells. Another example of a technological revolution is UWB (Ultra Wideband) technology, which, in principle, has been around since the 1950s, when communication and radio-localization systems using it were developed for military applications.

Recent advances in the design of miniature antennas and the performance of semiconductors allow us to consider the development of very high speed wireless systems based on this technology. UWB emissions are made up of ultra-short pulses (in the nanosecond range) occupying a very wide spectral field, from 500 MHz to a few GHz. These features make the UWB very attractive for developing wireless applications over short distances, in the range of a few meters, with a reduced power level. The main UWB applications fall under consumer communication systems (substituting wiring, multimedia interfaces for cell phones, etc.), and professional (electronic tags, imaging and detection systems) and security-related systems (short-range automotive radar for obstacle detection). As far as spectrum management is concerned, the conditions for the introduction of UWB systems are exclusively related to their interference potential. Identifying a future frequency allocation is made more difficult and complex by the very broad spectrum needing to be shared with numerous very different radio communication services (scientific services, space services, etc.).

Figure 1.1 presents the total dB attenuation, depending on the frequency of the electromagnetic wave towards the zenith on a clear sky. For local sizing of antennas for land–satellite links, statistical weather data as a percentage of time based on ITU recommendations should be taken into account. Depending on the frequency used, these attenuations cause an increase in the brightness temperature (noise temperature) of the antenna at reception, resulting in a sometimes significant degradation of the carrier-to-noise (C/N) ratio. For oblique and terrestrial connections, the higher the frequency, the more systematically the attenuation due to the crossing of the atmosphere must be taken into account in the link budget.

1.1.1. *The International Union of Telecommunications and Radiocommunications (ITU-R)*

Since 1865, the International Telecommunications Union (ITU) has been at the center of advances in communication, from telegraphy to the modern world of satellites, mobile phones, and the Internet. The ITU-R (formerly CCIR) recommendations are a set of international technical standards developed by ITU's radiocommunications sector. Today, these recommendations represent a report of nearly 5000 pages of information on how the limited resource of radio frequency spectrum and satellite orbits should be shared and used at the international level. They are the result of collaborative studies by ITU member states through radiocommunications study groups on:

– managing the spectrum of radio frequencies and orbits of geostationary or scrolling satellites;

– the use of a wide range of wireless services, including new mobile communications technologies;

– the efficient use of the radio frequency spectrum by all radiocommunications services;

– the propagation of radio waves;

– systems and networks of fixed satellite services, fixed and mobile services;

– space operations, satellite Earth-exploration, satellite weather and radio astronomy services;

– terrestrial and satellite broadcasting;

– spectrum management between countries and continents.

ITU member states endorse ITU-R recommendations, drawn up by experts from administrations, operators, industry, and other organizations dealing with radiocommunications issues around the world. Although they are not formally mandatory, the recommendations are followed around the world for technical and commercial reasons of compatibility between the services and between neighboring countries. However, regulation evolves over time and space: in space, according to the specifics of each country, and over time, according to technological innovations. Innovation is one of the main ways to gain a competitive advantage by meeting the needs of the market. However, regulation and innovation seem to contradict each other, since regulation, like all rules, is intended to be stable over time while innovation is unpredictable in principle. Effective regulation results in a trade-off between the protection and encouragement of innovators, which rewards innovation, and limits on anti-competitive behavior. Thus, ITU-R's work is essential, as it paves the way for the development of telecommunications through the introduction of new and expanded services, which, in turn, generate capacity and increased demand for new technologies. Broadband wireless access systems and other short-range devices (SRD) and ultra-wideband systems (UWB) are all examples of high-tech products and services based on ITU-R recommendations. Here, the antennas and the miniaturized sensors, integrated as close as possible to the active components, have a very important place when it comes to optimizing the capacity of a network.

In addition, the electromagnetic spectrum is regulated in different regions of the world by national or international organizations, such as ETSI[4] in Europe or the FCC[5] in the United States. This regulation includes the conditions for the use of different frequency bands, including whether or not there is a license to use them. For a licensed frequency band, an authorization is required and costs are charged to the user, as for mobile phone bands. Conversely, an unlicensed band, such as Wi-Fi bands, is freely usable by everyone within the limit of defined emission power levels.

1.1.2. *Frequency bands: uses and classification[6] (see also appendices 3 and 5)*

For radio communications, the spectrum of usable frequencies ranges from 3 Hz to 300 GHz. The usable portion of the spectrum is actually segmented into standardized bands with widths that vary as needed. ITU defines twelve standardized frequency bands. These bands are assigned, as part of a frequency band allocation table, to the various radiocommunication services, either exclusively or in a shared manner (Table 1.1). Any changes to this table (or the Radiocommunications Regulations, of which it is a part) can only be made by a world radiocommunications conference and only as a result of decisions adopted by consensus, or vote if necessary, through negotiations between the delegations of countries.

---

4 European Telecommunications Standards Institute (ETSI) is the European standardization organization in the telecommunications field. It is a non-profit organization whose role is to produce telecommunications standards for the present and the future.

5 The Federal Communications Commission (FCC) is an independent agency of the United States government created by the United States Congress in 1934. It is responsible for regulating telecommunications as well as the content of radio, television, and Internet broadcasts. Most commission officers (commissioners) are appointed by the President of the United States.

6 ITU-R, Article S2.1.

More and more short-range applications are emerging in the 100 THz and higher optical bands, where, for example, laser communications research is active. Table 1.1 provides an overview of this in terms of usage and classification. Wavelengths are calculated with the speed of light in the vacuum of $c = 3.108 \text{ m.s}^{-1}$. It was, by exact definition, set at $299{,}792{,}458 \text{ m.s}^{-1}$ in 1983 by the International Bureau of Weights and Measures (BIPM) in Paris.

The Radiocommunications Regulations (ITU-R 2016) state the following definition: "Radio waves (or hertzian waves): Electromagnetic waves of frequencies arbitrarily lower than 3000 GHz, propagated in space without artificial guide; they range from 9 kHz to 300 GHz, which corresponds to wavelengths of 33 km to 1 mm."

Frequency waves of less than 9 kHz are radio waves, but are not regulated. Frequency waves greater than 300 GHz are classified under infrared waves because the technology associated with their use is currently optical and non-electrical.

The range of 300 MHz to 300 GHz is considered the field of hyperfrequencies (microwaves), characterized by the fact that the circuits and equipment used are comparable in size to the wavelength.

Nomenclature	Frequency range (GHz)
VHF	0.03 – 0.30
UHF	0.30 – 1.00
L-band	1 – 2
S-Band	2 – 4
C-Band	4 – 8
X-Band	8 – 12
Ku-Band	12 – 18
K-Band	18 – 26.5
Ka-Band	26.5 – 40
Q-Band	33 – 50
U-Band	40 – 60
V-Band	50 – 75
E-Band	60 – 90
W-Band	75 – 110
F-Band	90 – 140
D-Band	110 – 170
G-Band	140 – 220

Table 1.2. *Nomenclature of IEEE fequency bands*

According to the ITU-R nomenclature, frequencies below 3 GHz are called *radio frequencies* (RF). Above that range, they are called *hyperfrequencies* (HF). Today, low frequency bands include the LF and MF frequencies from 3 kHz to 3 MHz.

1.1.3. *Review of some technologies by frequency bands*[7] *(see also appendices 3 and 5)*

The technology of an antenna depends mainly on the type of satellite or terrestrial application, the maximum power tolerated by the chosen technology, the link budget which depends on the sensitivity of the receiver, and the type of signal modulation.

1.1.3.1. *Low and medium frequencies (LF and MF) band*

Low and medium frequencies (LF and MF) ranging between 30 kHz and 3 MHz originally allowed ships to communicate with land stations and with each other, since these waves have the advantage of propagating over very long distances on the surface of the world. Likewise, they can penetrate deep into salt water, providing the possibility of communicating with submarines. However, at these very low frequencies, the bandwidth is very narrow. Consequently, the transmission of information was very limited, originally limited to telegraph signals. Radiotelephony and broadcasting only become possible in the upper part of the low frequency range. A major weakness relates to the nature of the equipment to be used. Indeed, these waves require, for their emission, very large energy generators and gigantic antennas made up of long wires, sometimes of several hundreds of meters stretched horizontally between pylons with, often, the ground as the ground plane. The problem for less "bulky" wired antennas is that it is necessary to increase the frequency, but this leads to a lesser range.

1.1.3.2. *High frequency (HF) band*

The high frequency (HF) band refers to frequencies between 3 MHz and 30 MHz. They are also called "short waves" or "decametric waves", that is to say, that their wavelength is between 10 and 100 meters. In the 1920s, the HF band was neglected in favor of lower frequencies whose propagation was considered more stable, but, as already mentioned, the LF and MF bands, sufficient for telegraphy, are no longer suitable for the transport of denser information. The range of HF waves is shorter in direct propagation but much longer by reflection on the ionosphere. The HF band is widely used in the least populated regions of the world, over longer distances than the VHF, especially since HF is barely disturbed by

7 A. Khaouja in LTE magazine editorial 2015.

natural and artificial obstacles. In the past, HF waves have been used to provide telephone and telegraph links between countries when the cable was lacking. By a double hop on the ionosphere, these connections are established from one side of the Earth to the other. In addition, the length of the radiating strands, constituting the HF antennas, makes the latter easy to deploy for fixed terrestrial applications, such as onboard ships. The energy consumed by the transmitter is much less than at low frequencies: a few kilowatts to communicate all around the world. Except for maritime and air services, the military and regions with poor infrastructure, HF waves are gradually being abandoned by official services and broadcasting due to the development of satellite and terrestrial relay links.

Figure 1.2. *HF telecommunication antenna (source: Wikipedia)*

1.1.3.3. *Very high frequency (VHF) band*

The very high frequency (VHF) band ranges from 30 MHz to 300 MHz for a wavelength ranging from 10 to 1 m. The frequencies in this band are mainly propagated in line-of-sight. Their wavelength is favorable to mobile or fixed links in radiotelephony with simple non-directional antennas and low power. The VHF band is shared among many uses. Terrestrial television broadcasting and FM (frequency modulation) broadcasting occupy half the spectrum, while the rest is allocated, in Europe, to satellite, maritime, amateur radio, aeronautical, private, military, or emergency services with portable transceivers. In most cases, the antennas are still wired.

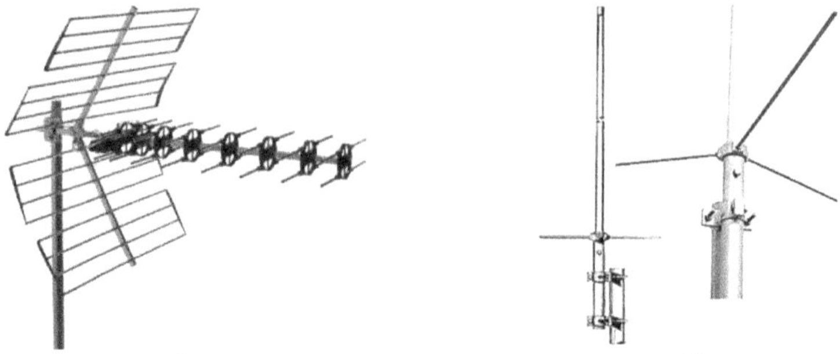

a) b)

Figure 1.3. *a) TV antenna; b) VHF/UHF antenna*

1.1.3.4. *Ultra-high frequency (UHF) band*

Officially, the ultra-high frequency (UHF) band is the band of the radio spectrum between 300 MHz and 3 GHz for a wavelength ranging from 1 m to 0.10 m. Frequencies above 1 GHz, up to around 300 GHz, are part of the technical domain of *hyperfrequencies* or *microwaves*, the range of which extends beyond that of decimeter waves to millimeter waves.

a)

b) c)

Figure 1.4. *a) Bluetooth dongle; b) omnidirectional Wi-Fi antenna; c) directional Wi-Fi antenna (2.4 GHz patch) (source: Wikipedia)*

UHF frequencies mainly have a line-of-sight propagation. Their use has grown with the saturation of the VHF band for fixed or mobile radiotelephony and television applications, as and when suitable components are available, such as MMIC technologies and, in particular, the shift from electronics to tubes with transistors. The band above 1 GHz was used first for radar, then for radio-relay systems, then for satellite links (GPS, mobile telephony), and then for GSM mobile phones and Wi-Fi or Bluetooth.

1.1.3.5. *Super high frequency (SHF) band*

The radio frequency band that ranges from 3 GHz to 30 GHz for a wavelength ranging from 0.10 m to 0.01 m is called super high frequency (SHF). SHFs are part of microwaves, which are used, in particular, in 2.45 GHz microwave ovens to agitate water molecules. However, these electromagnetic waves are very easily absorbed by any solid material of a certain thickness, which is not always an implementation advantage.

Figure 1.5. *Reflector antennas and new embedded technologies*

The frequencies from 3 GHz to 12 GHz allow connections between large urban centers, because the bandwidths allow for carrying a colossal amount of information. In fact, the carrier of a single transmitter can be modulated by thousands of telephone conversations and television signals. Furthermore, the high directivity of the antennas makes it possible to concentrate the waves in one or more narrow beams in one or more precise directions. As a result, all countries have set up a whole network of radio-relay systems that connect the main urban agglomerations. Likewise, high directivity antennas make it possible to connect all continents via satellite.

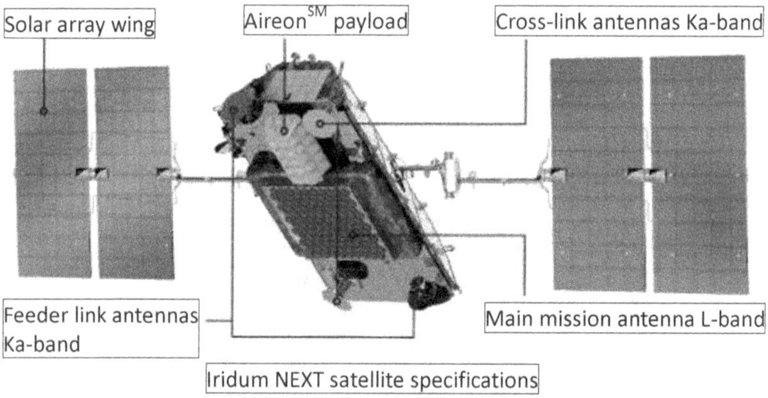

Figure 1.6. *Reflector antennas and new embedded technologies (continued)*

1.1.3.6. *Extremely high frequency (EHF) band*

The extremely high frequency (EHF) band ranges from 30 to 300 GHz, for a wavelength ranging from 0.01 m to 0.001 m. The frequencies in this band are also part of the microwaves. Today, millimeter frequency bands arouse great interest in radiocommunication systems because of the wide spectral bands available, that allow for very high transmission rates. The 60 GHz band (57–66 GHz) can be used without a license worldwide, which is a factor that is of particular interest for consumer applications. However, wireless communications at 60 GHz are limited by the resonance of air oxygen (Figure 1.1) which absorbs almost 98% of the energy emitted by the system, thus greatly limiting the communication range to a few meters, in the case of quasi-omnidirectional systems, and a few hundred meters, in the case of very directive systems.

These very high speed wireless communications systems are considered to be a very attractive technology for users of communicating objects (smartphones, tablets, laptops, set-top boxes, wireless hard drives, etc.) for transmission and synchronization of large amounts of data. Consequently, highly–integrated communication modules with low energy consumption are necessary to equip these devices, which that must be compact and light.

In this frequency band, the short wavelength makes it possible to integrate the antennas on the transmitter–receiver circuit. Two main integration approaches are possible for the integration of antennas. The SiP (System in Package) approach involves integrating a certain number of circuits from different technologies in a single package or module. This approach is expensive for consumer applications, and the various interconnections inherent in this technology are a major drawback at millimeter frequencies. The other approach, which is the SoC (System on Chip) approach, makes it possible to integrate both passive and active circuits on the same chip with the same technology. This approach is more promising and less expensive, and interconnection problems are less of a drawback with this approach than with the SiP approach (Dussopt 2012). The design of these antennas requires precise modeling of the integrated circuit, its case, and its immediate environment for good impedance control and the efficiency of the antenna. Today, the evolution of wireless technology in the millimeter band allows for speeds of several gigabits per second (Gb/s). Applications are available in different ranges, depending on the communication distance. For very short range systems, simple low-gain antennas are sufficient. Certain applications, with a range of ten meters or more, require antenna structures with high gain, good efficiency, and electronic beam steering.

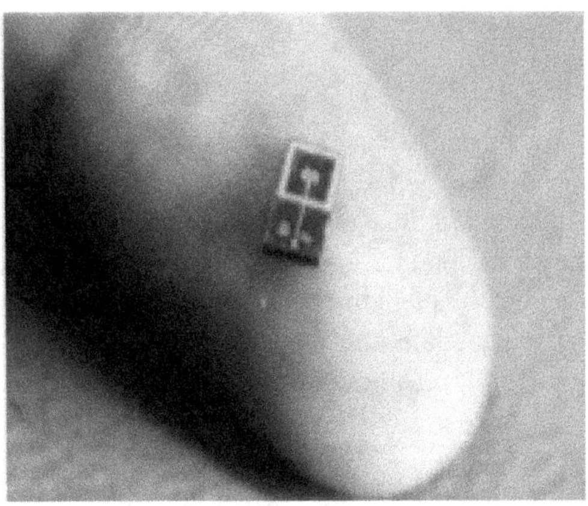

Figure 1.7. *Dipole antenna at millimeter frequencies (Houssemeddine* et al. *2013)*

The attenuation in free space between a transmitter and a receiver is given by equation [1.1]:

$$A(d) = -20\log_{10}\left(\frac{\lambda}{4.\pi.d}\right) \qquad [1.1]$$

This attenuation is already considerable at 60 GHz. It is 68 dB between a receiver and a transmitter separated by 1 m and 88 dB at 10 m. To this attenuation in free space the real conditions of propagation of the environment are added, such as the interior environment (indoors), where the signal encounters obstacles (walls, furniture, people, etc.), causing interference, which leads to fading with masking effects (shadowing), reflections or diffractions that considerably degrade the attenuation. Likewise, for losses, the higher the frequency, the more important the quality (homogeneity) of the materials, as well as the precision of the workmanship of the conductors and the quality of their surface (roughness). The size constraint and the losses generated in the silicium substrate, for example, lead to poor performance in terms of gain and efficiency.

Among other applications of the millimeter band, which is the subject of intense technological and theoretical research, there are various collision avoidance devices to improve road safety in cars.

Miniaturization allows for installing antennas for communication in the human body, which receives considerable attention from the scientific community, in terms of research, at the same time as the study of exposure to electromagnetic radiation on the human body.

1.1.3.7. *Terahertz (THz) band*

The terahertz (THz) band defined by the ITU ranges from 0.3 terahertz to 3 terahertz (1 THz = 10^{12} Hz), i.e. wavelengths ranging from 1 mm to 0.1 mm (100 μm). Because radiation in the terahertz band begins at a wavelength of 1 mm, it is also known as the submillimeter band, especially in astronomy. In the terahertz band, the energy of a photon is less than the band gap of non-metallic materials and, therefore, radiation can penetrate into these materials and be an alternative, in the medical field (non-ionizing) to using X-rays to obtain high-resolution images inside solid objects.

Terahertz radiation is located between infrared radiation and microwave radiation in the electromagnetic spectrum and shares common properties such as line-of-sight propagation and being non-ionizing. Like microwave radiation, it can propagate through a wide variety of non-conductive materials and its free space attenuation is considerably higher and sensitive to the environment (fog, cloud, etc.) and cannot penetrate water or metal (Choi 2010).

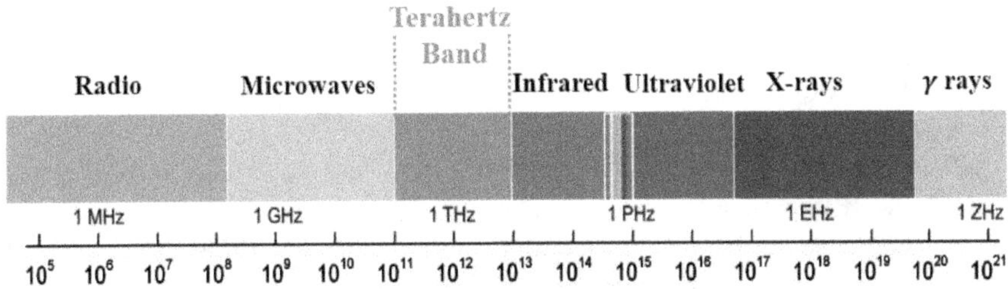

Figure 1.8. *Terahertz band*

As the terrestrial atmosphere is a strong attenuator of terahertz radiation (Figure 1.1), like the attenuation in free space (equation [1.1]), this limits its applications to a range of ten meters for broadband wireless networks, especially indoors.

While high frequencies facilitate the design of antennas and circuits of a size of the order of the wavelength, they nevertheless present difficulties linked to the high frequencies, such as the losses in conductors by the skin effect, losses in magnetic circuits and dielectrics, drop in gain of semiconductor components with frequency, parasitic elements in circuits (capacitances, inductances, couplings), and the difficulty of measuring the parameters of the diffraction S-matrix.

1.2. Propagation and radiation

The term "propagation" is reserved for, and used in, the case where the wave routing is guided by a material or natural structure (cable, waveguide, optical fiber, atmospheric layer, etc.) that defines a preferred direction in a waveguide or in space, as is the case of tropospheric propagation. The term "radiation" is used in the case of an emission for a free propagation in space, which is often the theoretically infinite "vacuum". However, many hybrid situations exist where the two terms, as in the field of antennas, have a fuzzy border. Therefore, the two terms will be adopted below: propagation in microwave circuits and radiation when the energy leaves the antenna.

1.3. Antenna and sensor

The antenna theory is fundamentally based on the Maxwell equations, elaborated on in Volume 1. In this second volume, we recall the crux of these equations, essential for understanding the functioning of an antenna and the theoretical data essential for the development of digital codes (software) for simulation.

To describe the performance of an antenna, parameters are defined according to an international IEEE standard (IEEE Standard 145-1993). Some of them are interdependent, and not all are essential to the performance description, but are useful, for example, within the framework of a technical specification when hyperfrequency sub-assemblies must be connected to it (amplifiers, filters, othomodes, etc.).

According to the IEEE definition, the antenna is a component of a device. It is well-positioned as the termination of a passive or active circuit. However, for the purpose of high integration, the antenna can be one of the components of an active circuit, such as a load, a capacitor, or even the battery or the case itself. In the case of a high integration, calculating the radiation performance must be done globally, with numerical methods adapted to the problem. More generally, the antenna is a component comprising metallic parts and/or linear and passive dielectrics. Active components can be added locally to obtain frequency, polarization, and radiation pattern agility. In biology, antennae are sensory (olfactory, tactile, etc.) organs that certain animal species, in particular, insects, have on their heads. These organs allow them to communicate with their environment.

A *sensor* is a device sensitive to a determined phenomenon and transforms this physical quantity into a signal (generally electrical: solar thermal collector, etc.). In the domain of space, a star sensor is an optoelectronic device used for the orientation of satellites. By this definition, "antenna" and "sensor" converge.

When the sensor is active, an active transmitting and/or receiving component is incorporated into it or associated with it, making it possible to recognize its environment (remote sensor). Active sensors pave the way for a technology that is the subject of intense research, called *smart skins*. The so-called "smart skin" sensor provides a new paradigm, where the surface of the structural object becomes an antenna, capable of understanding and adapting to changes in the physical parameters of its environment.

We will not distinguish between the terms *antenna* and *sensor* below, knowing that the former is used in the field of radiocommunication and the latter more systematically in the field of human sciences, in biology, medicine, in the field of earth sciences, etc.

Numerous works, such as (Rudge 1982; Silver 1984; James 1989; Milligan 2005; Balanis 2008) and many others, rightly considered as reference works, provide a detailed description of the characteristics of generic reference antennas, as well as wired antennas, aperture antennas, or antennas in planar technology (Figure 1.8). In each category, an antenna can have all shapes and sizes possible, such as fractal antennas (Felber 2000). The technology of the antenna depends on what one wants to do with it, ranging from space applications in the fixed or terrestrial domain for applications with mobiles right up to the connection between a laptop and a printer or a biological sensor, etc. An antenna consists almost exclusively of a metal part with high conductivity σ (siemens/m) and a part made of homogeneous dielectric materials, commonly of low relative permittivity ε_r and with low loss angle $tg(\delta)$.

In fact, we conventionally have three categories of technologies: wire antennas, aperture antennas, and antennas in printed technology, as shown in Figure 1.8, for example. Recent technologies add to these, such as the so-called *left hand* technologies, referencing the *right-hand* rule in electromagnetism. There are metamaterials, which are artificial materials with electromagnetic properties that do not exist in nature. These are also known as materials with a negative refractive index (Figure 1.9). The term *left hand materials* was coined by the Russian theorist Victor Georgievich Veselago, in 1968. In general, these are periodic, dielectric, or metallic structures which behave like a homogeneous material that does not exist in a natural environment. There are several types of metamaterials in electromagnetism, the best known being those capable of exhibiting both negative permeability and permittivity. But there also are others: media with infinite impedance, media with relative permittivity less than 1, etc. (Balanis 2008; Hao 2009; Andrei 2015; Lavrinenko 2015). Similarly, we can consider *photonic crystals* as metamaterials, but these can exist in their natural state, such as the scales of the wings of the *Morpho godarti* (Godart's morpho) butterfly, which are blue in clean air (refractive index $n = 1$), green in acetone vapor ($n = 1.36$), and light brown in trichloroethylene vapor ($n = 1.48$). The internal structure of the wing, interposing solid layers and layers of air, causes the vapors to change the optical index of the air layers, which modifies the spectrum reflected by the wing, like certain crystals that filter light according to permitted or prohibited optical frequency bands (PBG for photonic band gaps), or the colored glass used in the stained-glass windows of cathedrals. Photonic crystals are periodic structures of dielectric, semiconductor, or metallo-dielectric materials modifying the propagation of electromagnetic waves by creating bands of authorized and prohibited energy as in a semiconductor crystal. Photonic crystals are nanostructures. In microwaves, they are also sometimes called photonic band gap (PBG) materials.

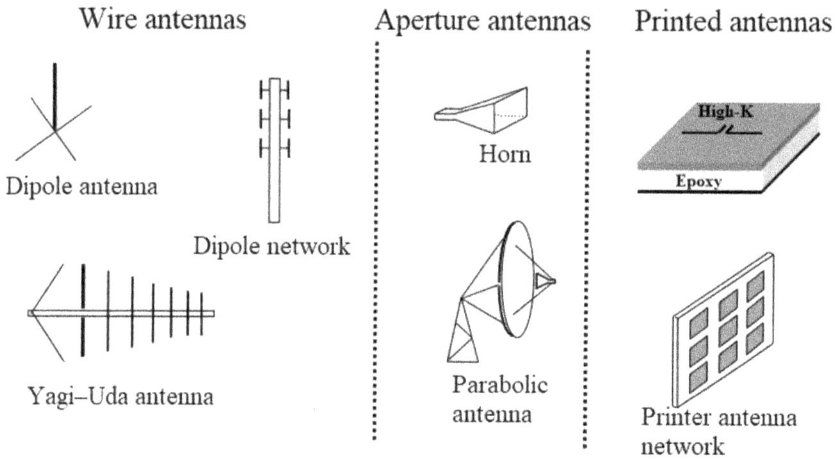

Figure 1.9. *Three major antenna technologies*

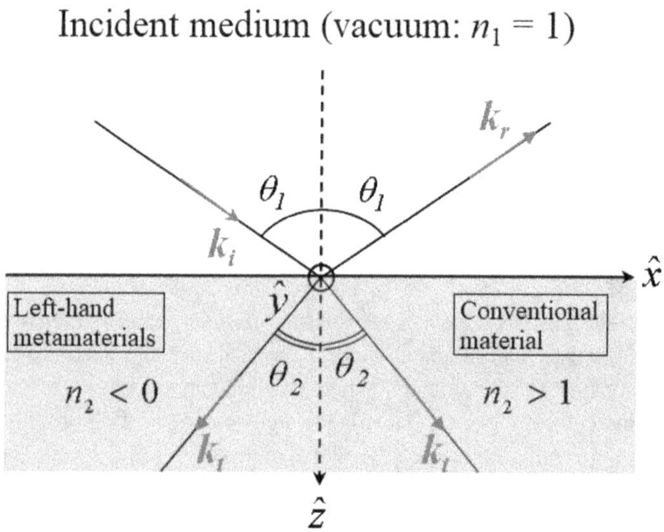

Figure 1.10. *Property of metamaterials (source: Wikipedia)*

a) b) c)

Figure 1.11. *Photonic crystals: a) 1D in one dimension; b) 2D in two dimensions; c) 3D in three dimensions (source: Wikipedia)*

A classification of antennas is based on the type of the radiating source. For certain antennas, the radiation source is made up of either linear or surface currents (dipole, loop, "planar" antenna, etc.). One could say that they radiate directly with free space, contrary to a horn on whose external surface the currents are prevented from circulating. To solve this problem, we calculate the electric and magnetic fields (or fictitious electric and magnetic currents) on a surface facing its "aperture".

1.3.1. *Antenna operating in transmission and reception*[7]

An antenna is made up of elements in which currents and volumetric or surface charges circulate, due to the radiation of energy, first guided by a transmission line connected either to an HPA (High Power Amplifier) transmitter or to an LNA (Low Noise Amplifier) receiver. This line, depending on the application and according to the frequency band (Table 1.1), is either a hollow metallic waveguide, filled with a dielectric or not, a coaxial cable, or a line in printed technology. Depending on the technology and, in particular, in the case of miniaturization, it is not always easy to precisely locate the reference plane a-b where the antenna "begins", which will always be calculated with a certain section of line and will then enable the connection and adaptation to the transmitter or receiver. The input impedance depends on the point where it is attacked on the line. This point must be chosen to have a maximum voltage or current in the radiating element. Giving an equivalent diagram of the antenna in the form of an electrical circuit, whether the constituent elements are located with R, L, C components (resistance, self, capacitance) or distributed (like line sections), is not always obvious, or even possible, according to its technology, its dimension in front of the wavelength, and the influence of the immediate environment that, by coupling, modifies the currents and, therefore, the impedance. In practice, it is the point where the reflection coefficient is measured

7 Silver 1984; Huang 2008.

that determines the position of the reference plane. In theory, the equivalent circuit will always assume that the calculated antenna is isolated in space outside the influence of the environment (case, human body, etc.) likely to modify its electrical and radioelectric characteristics. At the reception, like at the emission, an antenna can be represented by an Thevenin or Norton equivalent circuit.

1.3.1.1. *Transmitting antenna*

In order to ensure radiation in free space, it is necessary for a device to generate a radiated wave. The role of the transmitting antenna is to transform the guided electromagnetic power coming from a generator into radiated power. In this sense, it is a transducer.

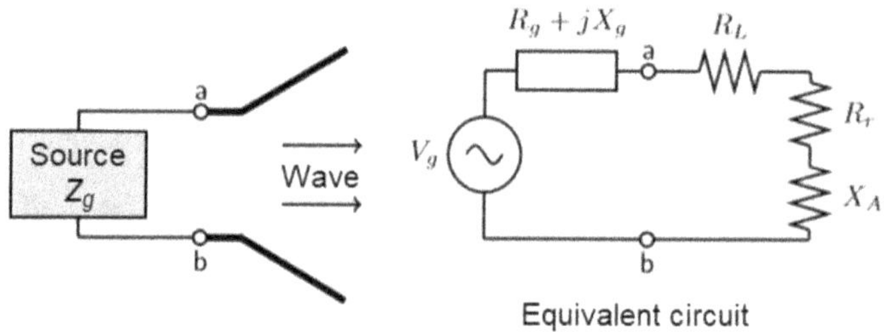

Figure 1.12. *Equivalent circuit in transmission (Thevenin)*

1.3.1.2. *Receiving antenna*

Conversely, the power radiated by an external source can be picked up by a receiving antenna. In this sense, the antenna appears as a sensor and a transformer of radiated power into guided electromagnetic power. A solar collector, like photovoltaic panels, can be considered as an antenna operating at very high frequency.

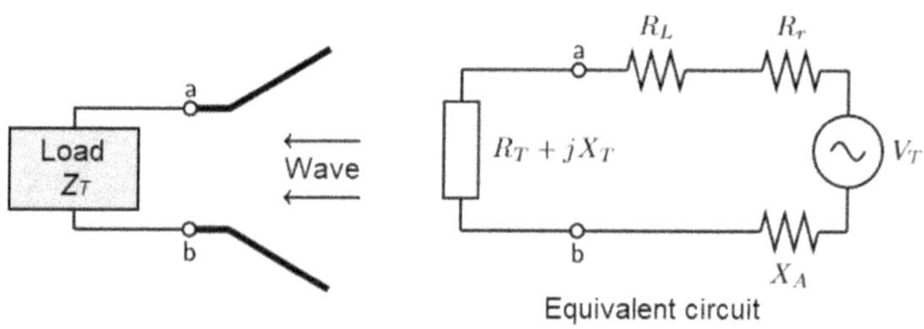

Figure 1.13. *Equivalent circuit in reception (Thevenin)*

With the input impedance at terminals a-b:

$$Z_A = R_A + jX_A \qquad [1.2]$$

where:

– $R_A = R_r + R_L$;

– R_r is the radiation resistance;

– R_L is the resistance of the antenna loss.

The radiation efficiency is:

$$\eta_r = \frac{R_r}{R_L + R_r}$$

The maximum power supplied to an antenna (radiated power) is 50% of the maximum power available through the generator when the latter is the complex conjugate of the antenna impedance, i.e. $Z_g = Z_A^*$ (Rudge 1982; Silver 1984; James 1989; Milligan 2005; Balanis 2008). In this case, the radiation efficiency, which is the ratio of the total radiated power to the power P_a available at the *a b* access of the antenna, is given by:

$$\eta_r = \frac{P_{radiated}}{P_a} = \frac{R_r}{(R_r + R_L)} = \frac{1}{\left(1 + {R_L}/{R_r}\right)} \qquad [1.3]$$

Equation [1.3] clearly shows that the lower the resistance, due to the different losses, and the higher the radiation resistance, the higher the efficiency of the antenna.

1.3.1.3. *Choice of adaptation point*

The essential part of the study of antennas is the development of methods of calculation and measurement of characteristic quantities, as defined by the IEEE standard 145-1993. To describe the performance of an antenna, a few parameters are necessary. Some are interdependent, but not all are necessarily specified for a complete description:

– impedance;

– polarization;

– radiation pattern;

– gain (directivity weighted by the efficiency);

– efficiency (all the causes of degradation of its yield, in particular, ohmic losses, unwanted radiation, etc.);

– the maximum power tolerated at transmission;

– mechanical bulk.

The antennas formed by a set of wired conductors were the first known and remain the most commonly used. They are a good starting point for the theoretical study of the general properties of antennas.

It should be noted that, except for the case of integrated active components whose impedance can exceed one hundred ohms (Ω), all equipment operating in RF, HF, and beyond has a standardized impedance of 50 Ω or 75 Ω. In all cases, the interconnections are always adapted in terms of impedance so as to provide maximum energy transfer. The complex impedance of an antenna varies with frequency. This corresponds to variations in the distribution of currents on its surface and/or in its volume. We always try to match the operating frequency with a purely real impedance point close to that of the system, i.e. 50 Ω, in general (Figure 1.13). However, the choice of the adaptation point, as indicated above, can determine the bandwidth.

The power range processed in RF (Table 1.1) goes from several tens of megawatts (MW) for broadcast transmitters to a few picowatts (pW) received, as is the case with a satellite telecommunications antenna, hence the interest in a perfect impedance adaptation. Given the possible power range, logarithmic scales are used and the power is expressed in decibels (dB), relative to a reference level of 1 milliwatt (mW) or 1 watt (see appendix 4):

$$P(dBm) = 10\log_{10}(P \text{ in } mW)$$

$$P(dBW) = 10\log_{10}(P \text{ in } W)$$

i.e.:

$$P(dBm) = P(dBW) + 30 \qquad [1.4]$$

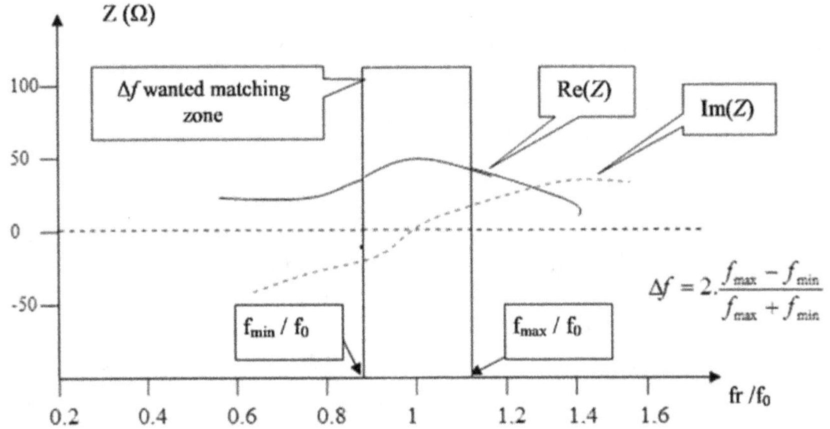

Figure 1.14. *The impedance adaptation point according to the bandwidth*

The advantage of a decibel rating is that the link budget of a transmission and reception chain can be done from start to finish by adding and subtracting the characteristics (amplification or losses) of each constituent element of the chain.

1.4. Theorems and important principles of electromagnetism[8]

The aim here is not to give or demonstrate all the theorems and principles, but to point out those that appear *a priori* to be essential during the theoretical and practical designing of a study.

1.4.1. *Lorentz reciprocity theorem*

This theorem considers the responses of a linear system to two different energizations and demonstrates a reciprocity between the responses. This theorem makes it possible to verify certain properties of diffracting objects, in particular, the reciprocity. Reciprocity results in the properties of the radiation from an antenna being the same in transmission and in reception. Thus, in most cases, an antenna can be used for reception or transmission with the same electrical and radioelectric properties, the same impedance, and the same radiation pattern. In some cases, when the antenna and/or the propagation medium include non-linear or anisotropic materials, they are not reciprocal. Because of the reciprocity, there will hardly ever

8 (Rudge 1982; Silver 1984; James 1989; Milligan 2005; Balanis 2008; Huang 2008).

be a difference between the radiation during transmission and reception. The qualities attributed to an antenna are in both operating modes if not specified.

1.4.2. *Huygens-Fresnel principle*

The aim of the equivalence principle is to replace the problem at hand with an equivalent problem, for which it is possible to find the solution more easily. More precisely, its principle is to replace the radiation of an antenna with that of simpler sources likely to be either modeled in an approximate manner or determined numerically. The principle is that the EM fields outside a volume V, delimited by a closed surface Σ, having a normal $\hat{n}$ directed towards the outside, surrounding one or more arbitrary sources located in V, can be obtained by substituting the initial sources with densities of electric $\vec{J}_s$ and magnetic surface currents $\vec{J}_{ms}$ on Σ. Magnetic currents have no physical existence, but their introduction into Maxwell's equations allows for simplifying certain problems. The fields inside V are, however, null after having eliminated the initial sources with Maxwell's equations. On Σ, the densities of surface currents are given by:

$$\vec{J}_S = \hat{n} \wedge \left[\vec{H}\right]_\Sigma \qquad [1.5]$$

$$\vec{J}_{mS} = \left[\vec{E}\right]_\Sigma \wedge \hat{n} \qquad [1.6]$$

where $\left[\vec{H}\right]_\Sigma$ and $\left[\vec{E}\right]_\Sigma$ are fields produced by the original sources evaluated on the surface Σ.

Equations [1.5] and [1.6] show that only the tangential components of the fields are involved.

This surface is arbitrary, but must be chosen according to the geometry of the antenna. Any closed surface is possible and all radiating sources can be substituted for it. In some cases, a portion of Σ is sufficient for a reliable, but not a rigorous result. This is the case when you know that the antenna is directive, like the so-called *aperture* antennas, or when the antenna is large, in order to limit the memory occupancy of the computer and/or the calculation time: parabolic antennas, open waveguides, horns, etc. For example, for the parabolic antenna in Figure 1.15a, the surface Σ is defined by the extreme rays coming from the primary source, reflected in the direction of the edge of the large reflector like an optical ray. In the case of small dimensions compared to the wavelength, as it is the case for the printed antenna in Figure 1.15b, there is a strong interest in keeping Σ closed. This theorem finds a systematic use in all numerical software like the FDTD (Finite Difference Time Domain) and the FEM (Finite Elements Method), for calculating the radiated fields at a great distance in free space. In these methods, in fact, we proceed to a discretization of the volume V and we calculate the EM field in the immediate area on Σ. The volume sampling is of the order of $\lambda/10$ to $\lambda/20$, which leads to a limited sampled volume, according to the power of the computer in "memory" space and in calculation time.

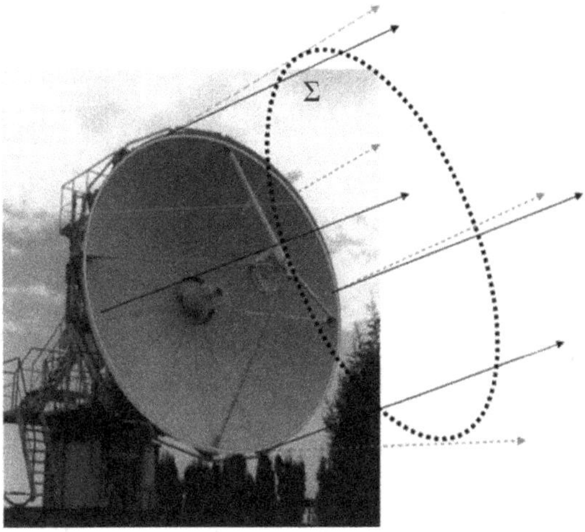

a)

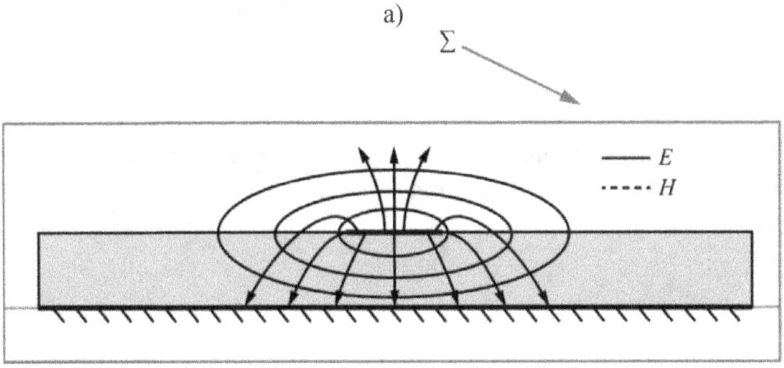

b)

Figure 1.15. *a) Equivalent area for a parabolic antenna; b) equivalent area for a printed antenna*

1.4.3. *Uniqueness theorem*

This theorem allows for explaining the conditions on the fields that make it possible to ensure that a solution is unique.

1.4.4. *Image theory*

The use of image charges is common in electromagnetism. This principle has been long-known in electrostatics. It is simply stated as finding an equivalence to the problem of a point load in the presence of a perfectly conducting infinite ground plane (PEC). There are two typical cases: a dipole parallel to the ground plane or a dipole perpendicular to the ground plane. Image theory allows for simplifying many softwares for simulating wire antennas.

1.4.5. *Superposition principle*

We will see, in section 2.1.2, which deals with material media, that if the medium is linear, homogeneous, isotropic, or non-isotropic, then the superposition principle applies, since Maxwell's equations are linear equations. Therefore, if a set of electric and magnetic vector fields $\vec{E}_1,.........$, $\vec{H}_1,.........$, and "source" functions $\rho_1,.........\ \vec{J}_1,.........$ satisfies Maxwell's equations and another set $\vec{E}_2,.........$, $\vec{H}_2,.........$, and $\rho_2,........$, $\vec{J}_2.........$ also satisfies them, then the sum of these solutions $\sum \vec{E}_i, \sum \vec{H}_i$, and $\sum \rho_i, \sum \vec{J}_i$ also satisfies Maxwell's equations. This principle is particularly important when, in the digital codes, the surfaces and the volumes are sampled in as many elementary sources of radiation, or when it is a question of the networking of elementary antennas, in addition to the translation theorem (Silver 1984).

2

Fundamental Equations Used in Antenna Design

Jean-Pierre BLOT
EUROSATCOM, Les Ulis, France

2.1. Formulations of Maxwell's equations to calculate the radiation of electromagnetic sources[1]

The principle upon which antennas work is based on the phenomenon of diffraction. To calculate their performances with a given technology, we use Maxwell's equations, which we saw in Volume 1 of this series. In view of the materials used to make antennas, here, we shall recap the main equations and formulations that must be used in devising numerical codes.

To begin with, all sources are rejected out of hand, and therefore, are not explicitly taken into account. Scientific advances have been made in recent times, using novel materials with a negative refractive index, like metamaterials known as "left-handed materials", or other structures which produce a gyrotropic effect. The development of these materials is still at the research stage, and they have complex formulations which are beyond our remit here to examine, given the complexity of the theory behind them. For an in-depth study of these media and the resulting technologies, readers may refer to the work of Vladimir V. Mitin (2017) and Sergei Tretyakov (2016), among others. Thus, our approach here will be a conventional one, looking at metal surfaces that are perfect electric conductors (PECs), the

For a color version of all figures in this book, see www.iste.co.uk/favennec/electromagnetic2.zip.

1 Rudge 1982; Silver 1984; Milligan 2005; Balanis 2008; Huang 2008; Picon 2009.

Electromagnetic Waves 2,
coordinated by Pierre-Noël FAVENNEC. © ISTE Ltd 2020

thickness of which is negligible in comparison to the wavelength λ. The dielectrics we discuss here will not be perfect insulators, but perform in a comparable way to the dielectrics most commonly employed in the field to build antennas and sensors: they will be linear, homogeneous, isotropic and non-magnetic.

2.1.1. *Maxwell's equations*

Step by step, we shall see in this section that Maxwell's equations contain all necessary data to design an antenna. The radiating components tend to be located in free space, but with certain technologies, they are surrounded by materials which may be complex and inhomogeneous, having a permittivity value of ε, permeability of μ and conductivity of σ; these properties may depend on the position, frequency and time, so it is prudent to retain a certain degree of general applicability in relation to free space. To recap what we saw in Volume 1, Maxwell's equations are expressed as follows:

– Maxwell–Gauss equation:

$$\vec{\nabla}.\vec{D}(x,y,z,t)=\rho(x,y,z,t) \tag{2.1a}$$

– Maxwell–Ampère equation:

$$\vec{\nabla}\wedge\vec{H}(x,y,z,t)=\vec{J}(x,y,z,t)+\frac{\partial\vec{D}(x,y,z,t)}{\partial t} \tag{2.1b}$$

– Maxwell–Faraday equation:

$$\vec{\nabla}\wedge\vec{E}(x,y,t)=-\frac{\partial\vec{B}(x,y,z,t)}{\partial t} \tag{2.1c}$$

– Maxwell–Thomson equation (magnetic flux equation):

$$\vec{\nabla}.\vec{B}(x,y,z,t)=0 \tag{2.1d}$$

The charge conservation equation, which derives from the above equations, is written as:

$$\vec{\nabla}.\vec{J}(x,y,z,t)=-\frac{\partial\rho(x,y,z,t)}{\partial t} \tag{2.1e}$$

$\vec{H}(A/m)$ is the magnetic field vector, measured in amperes per meter.

$\vec{B}(N/A.m)$ is the magnetic induction field vector, measured in tesla (N/amperes per meter).

$\vec{E}(V/m)$ is the electric field vector, measured in volts per meter.

$\vec{D}\left(C/m^2\right)$ is the electrical induction vector, measured in coulombs per square meter.

$\vec{J}(Amperes/m^2)$ is the electrical current density vector, measured in amperes per square meter.

$\rho\left(C/m^3\right)$ is the electrical charge density, measured in coulombs per cubic meter.

Readers may refer back to Volume 1 of this series for the meaning of each equation.

To simplify these formulae and make it easier to express the problem, we shall deal with the so-called "right-handed" Cartesian coordinate system, shown in Figure 2.1. A point P in space is defined by:

$$|\vec{r}| = \sqrt{x^2 + y^2 + z^2}\ ,\ \hat{x} \wedge \hat{y} = \hat{z}\ ,\ \hat{y} \wedge \hat{z} = \hat{x}\ ,\ \hat{z} \wedge \hat{x} = \hat{y}$$

In order to simplify the formulation, we set:

$$\vec{E}(x,y,z,t) = \vec{E}\ ,\ \vec{H}(x,y,z,t) = \vec{H}\ ,\ \vec{J}(x,y,z,t) = \vec{J}\ ,\ \rho(x,y,z,t) = \rho$$

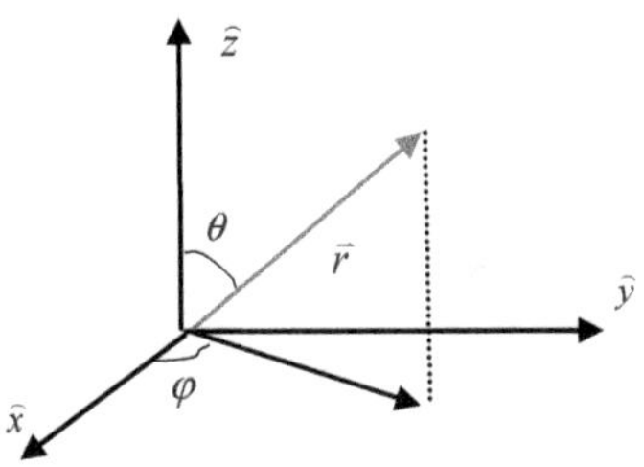

Figure 2.1. *Coordinate system*

2.1.2. *Material media*[2]

Maxwell's equations include three parameters that are characteristic of the materials and the material medium used to develop an antenna and its power supply circuit: conductivities σ (siemens/m), permittivities ε (farad/meter or coulomb/volt/meter) and permeabilities μ (henry/meter). In the field of antennas, the dielectrics must have very specific properties in order to optimize the effectiveness. They must be linear, homogeneous and isotropic (LHI). However, with modern computers, it is possible to take account of parameters which are not perfectly LHI, describing a material medium which is as close as possible to the physical and environmental reality.

2.1.2.1. *Polarization in insulating media*

In a conductive medium, the free charges or conduction charges (electrons or ions) can move freely throughout the material. This displacement gives us the free charge volume density ρ_{free} (C/m^3) and free current volume density $\vec{j}_{free}$ (A/m^2), as found in Maxwell's equations.

In a so-called *dielectric* insulating medium, there are no free charges, but bound charges can nevertheless move to some degree. Such displacement may give rise to induced dipole moments.

Thus, when it is subjected to an electrical field, a dielectric medium becomes polarized: each mesoscopic volume of matter $d\tau$ acquires an electric dipole moment $d\vec{p}$, induced by an electrical field $\vec{E}_{source}$, which may be applied by a source situated an infinite distance away. This dipole moment is characterized by a dipole moment per unit volume $\vec{P}$, known as the polarization vector, and defined by $d\vec{p} = \vec{P}.d\tau$. Where the stimulating field is weak, $\vec{P}$ is proportional to that field. When examining the electromagnetic field in a material medium at macroscopic scale, then for the polarization $\vec{P}$ of the medium in Maxwell's equations (see Volume 1, Chapter 1), we can substitute in:

– a volume polarization charge density $\rho_{pol} = -\vec{\nabla}.\vec{P}$;

– a surface polarization charge density $\sigma_{pol} = \overline{P}.\hat{n}$ (with $\hat{n}$ being orientated toward the outside of the surface of the volume of the material medium);

2 Rudge 1982; Silver 1984; James 1989; Milligan 2005.

– a volume polarization current density $\vec{J}_{pol} = \frac{d\vec{P}}{dt}$, under time-variable conditions.

In a material medium, we can still use Maxwell's equations, provided we take account of these different charge and current density values.

The volume polarization charge density ρ_{pol} and volume polarization current density $\vec{J}_{pol}$ satisfy the local charge conservation equation (see equation [1.202] in Volume 1):

$$\vec{\nabla}.\vec{J}_{pol} + \frac{\partial \rho_{pol}}{\partial t} = 0$$

We have the total charge $\rho_{total} = \rho_{free} + \rho_{pol}$ and the total current density $\vec{J}_{total} = \vec{J}_{free} + \vec{J}_{pol}$ (see equation [1.201b] in Volume 1).

2.1.2.2. *Magnetization in a material medium*

When subjected to a magnetic field $\vec{H}\ (A/m)$, a medium becomes magnetized: each mesoscopic volume of matter $d\tau$ acquires a magnetic dipole moment $d\vec{M}$ induced by the field, characterized by a volume dipole moment $\vec{M}$, known as the magnetization vector, and defined by $d\vec{M} = \vec{M}.d\tau$.

Certain media, known as ferromagnetic media, may be permanently magnetized.

When studying the electromagnetic field in a material medium, we can substitute the magnetization $\vec{M}$ of the medium with the following:

– a volume magnetization current density $\vec{J}_{mag} = \vec{\nabla} \wedge \vec{M}$;

– a surface magnetization current density $\vec{J}_{Smag} = \vec{M} \wedge \hat{n}$ (with $\hat{n}$ being the normal, orientated toward the outside of the material medium).

In a material medium, we can continue to use the expressions of Maxwell's equations formulated to express behavior "in a vacuum", provided we take account of the magnetization current volume density, written as J_{mag} for magnetic current density.

2.1.3. Vectors $\vec{D}$ and $\vec{H}$

Our aim is to be able to use equations similar to those which operate in a vacuum. In order to do so, we introduce two new vectors:

– $\vec{D} = \varepsilon_0 \vec{E} + \vec{P}$, the electric induction vector or displacement vector (see Volume 1, Chapter 1);

– $\vec{H} = \dfrac{\vec{B}}{\mu_0} - \vec{M}$, the magnetic excitation vector (see equation [1.235], Volume 1, Chapter 1).

These two relations are known as *constitutive equations*, because they form the connection between the material's response and the excitation to which it is subjected, be it an electrical field or a magnetic field.

In addition to the presence of a magnetic current distribution $\vec{J}_{mag}$ in diffraction theory, an equivalent magnetic charge ρ_{mag} is introduced, to make Maxwell's equations more symmetrical. It should be noted that, to date, no physical evidence of the existence of $\vec{J}_{mag}$ and ρ_{mag} has been found. This is surprising, because the behavior of the electrical field and of the magnetic field is otherwise relatively similar. Constructing parameters similar to the charge density and electrical current in relation to a magnetic field aids the symmetry of the equations. One further term is added to the Maxwell–Thomson equation, and another to the Maxwell–Faraday equation. The search for magnetic monopoles is the subject of a great deal of research in quantum mechanics. Using the results from Tables 1.6 and 1.7 of Volume 1, and integrating the magnetic current density vector $\vec{J}_{mag}$ and the equivalent magnetic charge density ρ_{mag}, Maxwell's equations in a material medium can be rewritten as follows:

– Maxwell–Gauss equation:

$$\vec{\nabla}.D = \rho_{free} \qquad [2.2a]$$

– Maxwell–Ampère equation:

$$\vec{\nabla} \wedge \vec{H} = \vec{J}_{free} + \frac{\partial \vec{D}}{\partial t} \qquad [2.2b]$$

– Maxwell–Faraday equation:

$$\vec{\nabla} \wedge \vec{E} = -\vec{J}_{mag} - \frac{\partial \vec{B}}{\partial t} \qquad [2.2c]$$

– Maxwell–Thomson equation:

$$\vec{\nabla}.\vec{B} = \rho_{mag} \qquad [2.2d]$$

We must also take account of the continuity equation for the electrical charges and currents:

$$\vec{\nabla}.\vec{J}_{free} + \frac{\partial \rho_{free}}{\partial t} = 0 \qquad [2.2e]$$

and the continuity equation for the magnetic charges and currents:

$$\vec{\nabla}.\vec{J}_{mag} + \frac{\partial \rho_{mag}}{\partial t} = 0 \qquad [2.2f]$$

where:

$$\vec{B} = \mu_0 \left(\vec{H} + \vec{M}\right) = \left(1 + \chi_m\right) \vec{H} = \mu_0 . \mu_r \vec{H}$$

where χ_m is the magnetic susceptibility. With this knowledge, we can write:

$$\vec{H} = \frac{1}{\mu_0 \mu_r} \vec{B} = \frac{1}{\mu} \vec{B} \qquad [2.3]$$

where μ (henry/meter) is the absolute magnetic permeability and μ_r the relative magnetic permeability (dimensionless).

NOTE.– In a complex, non-LHI medium, $\vec{H} = g\left(\vec{B}\right) = [\mu]^{-1}.\vec{B}$, where g is a function, $[\mu]^{-1}$ is a third-order tensor – i.e. a 3 × 3 matrix with complex or real coefficients, having an absolute magnetic permeability of $\mu_{i,j} = \mu_0 \mu_{i,jr}$.

In an LHI medium, the function g is linear, and there is no prevailing region or direction. Hence, the matrix can be condensed to a scalar.

2.1.3.1. *Local Maxwell equations in a material medium*

In a medium where $\varepsilon = \varepsilon_0$ and $\mu = \mu_0$, using $\rho_{pol} = -\vec{\nabla}.\vec{P}$, $\vec{J}_{pol} = \frac{\partial \vec{P}}{\partial t}$ and $\vec{J}_{mag} = \vec{\nabla} \wedge \vec{M}$, Maxwell's equations are expressed as follows:

– Maxwell–Gauss equation:

$$\vec{\nabla}.\vec{E} = \frac{\rho_{free} + \rho_{pol}}{\varepsilon_0} = \frac{\rho_{free}}{\varepsilon_0} - \frac{\vec{\nabla}.\vec{P}}{\varepsilon_0} \qquad [2.4a]$$

– Maxwell–Ampère equation:

$$\vec{\nabla} \wedge \vec{B} = \mu_0 \left(\vec{J}_{free} + \frac{\partial \vec{P}}{\partial t} + \vec{\nabla} \wedge \vec{M} \right) + \varepsilon_0 \mu_0 \frac{\partial \vec{E}}{\partial t} \qquad [2.4b]$$

– Maxwell–Faraday equation:

$$\vec{\nabla} \wedge \vec{E} = -\frac{\partial \vec{B}}{\partial t} \qquad [2.4c]$$

– Maxwell–Thomson equation (magnetic flux equation):

$$\vec{\nabla}.\vec{B} = 0 \qquad [2.4d]$$

These equations are not sufficient to determine the electromagnetic field $\left(\vec{E}, \vec{B}\right)$, because the polarization $\vec{P}$ and magnetization $\vec{M}$ functions are generally not known in advance, but rather depend, themselves, on $\vec{E}$ and $\vec{B}$.

2.1.3.2. *Linear, homogeneous and isotropic (LHI) media*[3]

A medium is said to be *linear* if $\vec{P}$ and $\vec{E}$ are bound by a relation of the type $\vec{P} = \varepsilon_0.\chi_e.\vec{E}$ when $\left|\vec{P}\right|$ is sufficiently small, where χ_e is a dimensionless positive number called the *dielectric susceptibility*. In a linear medium, the permittivity ε and permeability μ are independent of the fields applied.

---

3 Rudge 1982; Silver 1984; James 1989; Milligan 2005.

A medium is said to be *isotropic* if its properties are independent of the *direction* in question, and where the permittivity and permeability are scalar values and $\vec{P} // \vec{E}$.

A medium is said to be *homogeneous* if the permittivity or permeability (which may be time-dependent) do not depend on the point in the space.

A material is said to be *perfect* if it is linear, homogeneous, isotropic and lossless. This is an approximation which, to begin with, helps simplify the equations to produce a quick design for devices. In an LHI medium, $\vec{E}$, $\vec{P}$, $\vec{D}$ are linked by the relations (see equations [1.222] and [1.223] in Volume 1):

$$\vec{P} = \varepsilon_0 . \chi_e . \vec{E} \qquad [2.5a]$$

$$\vec{D} = \varepsilon_0 (1 + \chi_e) \vec{E} = \varepsilon_0 \varepsilon_r \vec{E} = \varepsilon \vec{E} \qquad [2.5b]$$

where $\varepsilon_r = 1 + \chi_e$ (dimensionless), ε_r (farad/meter) the relative permittivity of the medium (see equation [1.224] in Volume 1).

The relation between $\vec{E}$ and $\vec{D}$, in the case of an LHI material, exhibiting an *instantaneous response* to changes in the electric field, is:

$$\vec{D} = \varepsilon \vec{E} \qquad [2.6]$$

where ε is positive and constant.

NOTE.– In a non-LHI complex medium, $\vec{D} = f(E) = [\varepsilon] \vec{E}$ where f is a function, and $[\varepsilon]$ a 3 × 3 matrix with complex or real coefficients $\varepsilon_{i,j}$. In an LHI medium, the function f is linear, there is no prevailing region, and no prevailing direction. Thus, the matrix can be condensed to a scalar.

In any medium whatsoever, if the material is not isotropic – that is, where ε is a matrix $[\varepsilon]$ – then the vector field $\vec{D}$ is not collinear with $\vec{E}$. If the material is not homogeneous, the coefficients $\varepsilon_{i,j}$ in the matrix $[\varepsilon]$ depend on the spatial coordinates x, y, z (see equation [1.217] in Volume 1). If the material does not have an instantaneous response, the coefficients $\varepsilon_{i,j}$ depend either on time t or on frequency $\omega = 2\pi.f$. In this case, $\varepsilon_r(\omega) = 1 + \chi_e(\omega)$ is a complex value. For example, the Drude model gives us a relative permittivity:

$$\varepsilon_r(\omega) = 1 - \frac{\omega_p^2}{\omega^2 - j.\omega.\Gamma_D} \qquad [2.7]$$

where:

$$j = \sqrt{-1}$$

$$j^2 = -1$$

$$\Gamma_D = \frac{1}{\tau} \text{ (damping term)}$$

τ is the mean time between two collisions or relaxation time from collisions between the electrons and ions in the material.

$\omega_p = \sqrt{\dfrac{N_c.e^2}{\varepsilon_0.m_e}}$ is the pulsation characterizing the collective oscillation of the conduction electrons in the material, known as *plasma* pulsation, where m_e is the mass of an electron and N_c the number of conduction electrons.

From equation [2.6], we can see that while the fields $\vec{E}$ and $\vec{D}$ are exerted in the same direction, they may not necessarily always be in phase with one another. These phase shifts depend on the molecular nature of the medium, and are linked to the loss of electromagnetic energy in the medium (dielectric losses). It is practical to express ε as a complex number:

$$\varepsilon = \varepsilon' - j\varepsilon'' \qquad [2.8]$$

The energy lost, associated with the imaginary part of ε, must be set apart from the conduction losses associated with the conduction currents $\vec{J}$. In the study of antennas, barring a few rare exceptions, the materials used are LHI, and non-magnetic materials with permeability μ_0. Only conduction currents exist in ohmic media, and are such that:

$$\vec{J}_{induced} = \sigma.\vec{E} \qquad [2.9]$$

where σ (siemens/meter) is the conductivity of the medium. Similarly to the other parameters, σ may be frequency-dependent.

A conductive medium cannot contain a volume density of free charges ρ; if the value of σ is high, the volume charges ρ are null ($\rho = 0$), and are not time-dependent. In addition, if all of these parameters (σ, ε and μ) are independent of the amplitude of the fields $\vec{E}$ and $\vec{H}$, Maxwell's equations are linear, including relations [2.6] and [2.9]. *Consequently, the superposition principle is applicable*: this means that the total field resulting from the sum of the fields $\vec{E}_i$ and $\vec{H}_i$ associated with the charges ρ_i and the densities $\vec{J}_i$ is such that $\Sigma\vec{E}_i$, $\Sigma\rho_i$ and $\Sigma\vec{J}_i$ also satisfy Maxwell's equations.

The materials used in antenna design exhibit very low magnetic properties, and are generally LHI. Very broadly speaking, we have:

$$\vec{M} = \chi_m.\vec{H} \text{ and } \vec{B} = \mu_0.(1+\chi_m).\vec{H} = \mu_0.\mu_r.\vec{H} = \mu.\vec{H}$$

Note that if χ_e and χ_m are complex and independent of the frequency, we see both dielectric and magnetic losses. If $\chi_e(\omega)$ and $\chi_m(\omega)$ are complex and frequency-dependent, then we are dealing with a dispersive medium.

From an electromagnetic standpoint (Kraus 1999; Huang 2008), materials may be classified as conductive, semi-conductive or dielectric. The properties of these materials, as we saw earlier, are generally dependent on the frequency, and losses are seen. They are characterized by a loss tangent $tg(\delta) = \frac{\varepsilon^{'}}{\varepsilon^{'''}}$:

– as a conductor: $tg(\delta) = \frac{\varepsilon^{'}}{\varepsilon^{'''}} = \frac{\sigma}{\omega.\varepsilon} > 100$;

– as a semiconductor: $0.01 < tg(\delta) = \frac{\sigma}{\omega.\varepsilon} < 100$;

– as a dielectric: $tg(\delta) = \frac{\varepsilon^{'}}{\varepsilon^{'''}} = \frac{\sigma}{\omega.\varepsilon} < 0.01$.

2.1.4. *Source currents and induced currents*[4]

When dealing with any problem involving electromagnetism, we accept the existence of source currents $\vec{J}_{source}$ which cannot be modified either by the fields that they generate or by any other field. Such source currents generate electromagnetic *excitation* fields. If any given object is placed in the vicinity of the sources, these excitation fields (incident fields) produce induced currents $\vec{J}_{induced}$ within the object.

In turn, these currents induced in the object generate *diffracted* fields. The total field is the sum of the excitation fields (incident) and the diffracted fields, such that $\vec{E}_{total} = \vec{E}_{incident} + \vec{E}_{diffracted}$, $\vec{H}_{total} = \vec{H}_{incident} + \vec{H}_{diffracted}$, created by these two types of currents $\vec{J}_{source}$ and $\vec{J}_{induced}$. There is no physical difference between these two types of currents – essentially, they are streams of electrons in motion. Nevertheless, we must establish a crucially important conceptual difference:

– $\vec{J}_{source}$ is a *known* and *imposed* current. It is not affected by the existing field. The source current *is the given* in the problem with which we are dealing;

– $\vec{J}_{induced}$ is a current which is usually *unknown*, which is *induced* in objects in the vicinity of the source, and depend on the total field $\vec{E}_{total}$.

In the Maxwell–Ampère equation $\vec{\nabla} \wedge \vec{H} = \vec{J} + j.\omega.\varepsilon.\vec{E}$, the current $\vec{J}$ is a total value, and therefore $\vec{J} = \vec{J}_{source} + \vec{J}_{induced}$, $\vec{E} = \vec{E}_{total}$ and $\vec{H} = \vec{H}_{total}$. Our aim is to isolate the portion of the total current which is independent of the fields. This is easy to do for objects made of linear, non-magnetic materials, because the induced current is linked solely to the total electric field, by way of Ohm's law: $\vec{J}_{induced} = \sigma.\vec{E}_{total}$.

Thus, we can write:

$$j.\omega.\varepsilon.\vec{E}_{total} + \vec{J} = j.\omega.\varepsilon.\vec{E}_{total} + \vec{J}_{source} + \vec{J}_{induced} = \left(j.\omega.\varepsilon + \sigma\right)\vec{E}_{total} + \vec{J}_{source}$$

$$= j.\omega.\left(\varepsilon + \frac{\sigma}{j.\omega}\right)\vec{E}_{total} + \vec{J}_{source}$$

4 Mosig 2007; Gerl 1970.

If we set:

$$\varepsilon_T = \varepsilon + \frac{\sigma}{j.\omega} = \varepsilon - j.\frac{\sigma}{\omega} = \varepsilon' - j.\varepsilon'' \qquad [2.10]$$

we can write:

$$\vec{\nabla} \wedge \vec{H}_{total} = j.\omega.\varepsilon_T.\vec{E}_{total} + \vec{J}_{source} \qquad [2.11]$$

However, for ease of writing, we shall continue to use the original form of the equation. From hereon in though, $\vec{J}$ will be such that $\vec{J} = \vec{J}_{total}$ with a complex ε (see equations [2.8] and [2.10]), which introduces a further negative imaginary part $-j.\frac{\sigma}{\omega}$ for LHI materials.

In conclusion, Maxwell's equations, for our purposes here in creating simulation software, are set forth in Box 2.1.

Maxwell–Gauss equation:

$$\vec{\nabla}.\vec{E} = \frac{\rho_{total}}{\varepsilon}$$

Maxwell–Ampère equation:

$$\vec{\nabla} \wedge \vec{B} = \mu(\vec{J}_{total} + \varepsilon.\frac{\partial \vec{E}}{\partial t})$$

Maxwell–Faraday equation:

$$\vec{\nabla} \wedge \vec{E} = -\vec{J}_{mag} - \frac{\partial \vec{B}}{\partial t} \;, \; \vec{E} = -\frac{\partial \vec{A}}{\partial t} - \vec{\nabla}.V$$

Maxwell–Thomson equation (magnetic flux equation):

$$\vec{\nabla}.\vec{B} = \rho_{mag} \;;\; \vec{B} = \vec{\nabla} \wedge \vec{A}$$

where $\vec{J}_{total} = \vec{J}_{source} + \vec{J}_{induced} + \vec{J}_{mag} + \vec{J}_{pol}$; $\rho_{total} = \rho_{source} + \rho_{pol}$.

In a non-magnetic, non-polarizable medium $\vec{J}_{total} = \vec{J}_{source} + \vec{J}_{induced}$ and $\rho_{total} = \rho_{source}$:

$$\vec{J}_{induced} = \sigma.\vec{E}$$

V and $\vec{A}$, respectively, are the scalar potential and the vector potential (section 3.9).

$$\vec{J}_{total} = \vec{J}_{source} + \vec{J}_{induced} = \vec{J}_{source} + \sigma.\vec{E}$$

$\vec{J}_{total}$ is the sum of an excitation current density $\vec{J}_{source}$ imposed by the experimenter and a current density $\vec{J}_{induced} = \sigma.\vec{E}$ induced by $\vec{E}$, which manifests as the response to the excitation current $\vec{J}_{source}$. Similarly, the total charge density $\rho_{total} = \rho_{induced} + \rho_{pol}$ where $\rho_{induced}$ is the charge density induced by the current $\vec{J}_{source}$.

Box 2.1. *Overview of equations for the development of a simulation code*

2.1.5. *Integral form of Maxwell's equation*[5]

Depending on the problem at hand, Maxwell's equations are used in the form of their local expressions (see above), as is done in FDTD, or in the integral form (see Volume 1, Chapter 1, section 1.2.4.1), as is used in the method of moments (MoM).

2.2. Boundary conditions between two media[6]

Maxwell's equations lead to certain constraints on electromagnetic fields in the presence of different media $(\varepsilon_n, \mu_n, \sigma_n)$. Consider two media $(\varepsilon_1, \mu_1, \sigma_1)$ and $(\varepsilon_2, \mu_2, \sigma_2)$ with a normal $\hat{n}$ at the separating surface, directed from medium 1 to medium 2. These media need not necessarily be LHI.

In order to establish the transition relations which take effect when we move between two media, we consider two points M_1 and M_2, both very near to the interface, and belonging, respectively, to medium 1 and medium 2. Each of

5 Rudge 1982; Silver 1984; James 1989; Milligan 2005; Balanis 2008.

6 Silver 1984; Milligan 2005; Balanis 2008; Huang 2008.

Maxwell's equations impose a transition condition. The demonstrations are carried out at infinitesimal dimensions. With this in mind, we can consider the interface to be locally planar, and the equations obtained to be local. The procedure is standard in numerous publications, based on the integral equations among others (Statton 1941; Picon 2009).

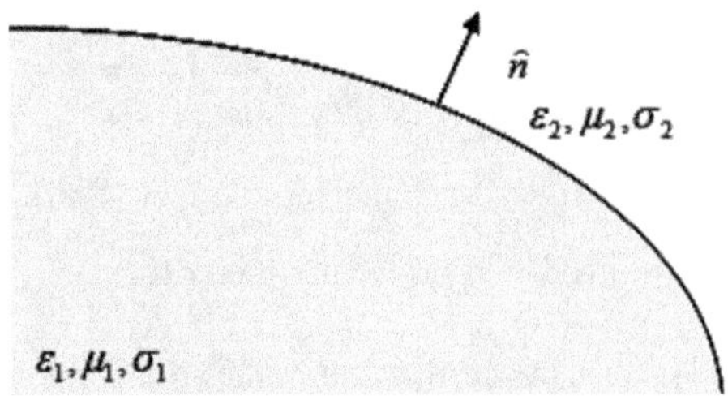

Figure 2.2. *Boundary conditions between two dielectric media*

Consider the vector fields $\vec{E}_1, \vec{H}_1$ in medium 1, infinitely close to the separating surface, and also $\vec{E}_2, \vec{H}_2$, in medium 2. The boundary conditions are presented below.

– *The tangential component of the electric field is continuous on crossing the boundary*:

$$\hat{n} \wedge \left(\vec{E}_2 - \vec{E}_1\right) = \vec{0} \quad [2.12]$$

It can be demonstrated that the field penetrates into a conductive medium to a depth δ which is inversely proportional to the square root (equation [2.13]) of the conductivity. For a good conductor $\left(\frac{\sigma}{\omega.\varepsilon} > 100\right)$:

$$\delta \approx \sqrt{\frac{2}{\omega.\mu.\sigma}} = \sqrt{\frac{2.\rho}{\omega.\mu}} = \frac{1}{\sqrt{\sigma.\mu.\pi.f}} \quad [2.13]$$

The *surface resistivity* of the conductor is given by:

$$R_s = \sqrt{\frac{\omega.\mu}{2.\sigma}} = \sqrt{\frac{\mu.\pi.f}{\sigma}} \quad [2.14]$$

where:

– δ is the thickness of the skin in meters [m];

– ω is the pulsation in radians per second [rad/s] (ω=2.π.f);

– f is the frequency in hertz [Hz];

– μ is the magnetic permeability in henry per meter [H/m];

– ρ is the resistivity in ohms per meter [Ω.m] ($\rho = 1/\sigma$);

– σ is the electrical conductivity in siemens per meter [S/m];

– ϵ is the dielectric permittivity in farads per meter [F/m].

From equation [2.14], we can see that the higher the frequency f, the higher the surface resistivity. At very high frequencies (particularly frequencies in the THz range), depending on the nature of the conductors, this can cause the problem of overall ohmic loss of the circuits, also caused by the rugosity of the surfaces.

If the conductivity is infinite, then ($\sigma_1 \rightarrow \infty$). This is only the case when medium 1 is a perfect electric conductor (PEC): $\vec{E}_1 = \vec{0}$, where $\vec{E}_2 = \vec{E}_{Total}$ = $(\vec{E}_{incident} + \vec{E}_{radiated})$, then:

$$\hat{n} \wedge \vec{E}_2 = \vec{0} \quad [2.15]$$

The model of a *perfect* conductor, in addition to LHI materials, is very widely used in antenna design. These assumptions make the calculations easier.

– *The tangential component of the magnetic field is discontinuous on crossing the surface only if a surface current density is present on the boundary:*

$$\hat{n} \wedge \left(\vec{H}_2 - \vec{H}_1\right) = \vec{J}_S \quad [2.16]$$

If medium 1 is a PEC:

$$\hat{n} \wedge \vec{H}_2 = \hat{n} \wedge H_{Total} = \hat{n} \wedge \left(\vec{H}_{incident} + \vec{H}_{radiated}\right) = \vec{J}_S \quad [2.17]$$

– *We see discontinuity of the normal component of* $\vec{D}$ *if there is a surface charge density* ρ_s *on the boundary:*

$$\hat{n}.\left(\vec{D}_2 - \vec{D}_1\right) = \rho_s \quad [2.18]$$

If medium 1 is a PEC:

$$\hat{n}.\vec{D}_2 = \hat{n}.\varepsilon_2 \vec{E}_{Total} = \rho_s \quad [2.19]$$

– The normal component of the magnetic field varies continuously on crossing the boundary:

$$\hat{\text{n}}.\left(\vec{B}_2 - \vec{B}_1\right) = \hat{n}(\mu_2.\vec{H}_2 - \mu_1.\vec{H}_1) = 0 \quad [2.20]$$

If medium 1 is a PEC:

$$n.\vec{B}_2 = n.\mu_2.\vec{H}_{Total} = 0 \quad [2.21]$$

If both media are perfect dielectrics $(\sigma_1 = \sigma_2 = 0)$, the current density and charge density at the surface are null: $\rho_e = 0$, $\vec{J}_s = \vec{0}$.

The boundary conditions apply universally to fields which satisfy Maxwell's equations. These discontinuities can formally associate fictitious magnetic charge distributions and currents ρ_{mag} and $\vec{J}_{mag}$ on the surface separating the two media. By the same reasoning, we can obtain the following formulae:

$$\hat{n} \wedge \left(\vec{E}_2 - \vec{E}_1\right) = -\vec{J}_{mag} \quad [2.22]$$

$$\hat{n}.\left(\vec{B}_2 - \vec{B}_1\right) = \rho_{mag} \quad [2.23]$$

where, on the surface, the conservation relation below applies:

$$\vec{\nabla}_S.\vec{J}_{mag} + \frac{\partial \rho_{mag}}{\partial t} = 0 \quad [2.24]$$

2.3. Vector potential[7]

To simplify the formulae and the presentation here, the propagation medium is assumed to be *homogeneous* and *isotropic*. In numerous cases in physics, based on

7 Gerl 1970; Huang 2008; Picon 2009.

the solving of differential equations, it is common to use auxiliary functions known as potentials. When solving Maxwell's equations, two potentials come into play:

– the vector potential, written as $\vec{A}$, also known as the *magnetic potential*;

– the scalar potential denoted by V, also known as the *electric potential.*

The electromagnetic (EM) field is presented as a real field, whereas the potentials appear as a mathematical tool, helping to solve a number of problems in the domain of antennas, because expressions which include them are simpler than they are with the electric field $\vec{E}$ and the magnetic induction field $\vec{B}$. Vectorial differentiation of these potentials gives us the electric and magnetic fields.

The vector potential of the magnetic field is usually introduced as the result of Maxwell's equations, which stipulate that the divergence of the magnetic field $\vec{B}$ must be zero. Vector analysis indicates that a three-dimensional vector field with zero divergence can always be expressed in the form of the rotational of a vector field, written as $\vec{A}$. Thus, we have:

$$\vec{\nabla}.\vec{B} = 0 \Rightarrow \vec{B} = \vec{\nabla} \wedge \vec{A} \qquad [2.25]$$

$\vec{A}$ is referred to as the vector potential of $\vec{B}$, while V is the *scalar* potential of $\vec{E}$.

In addition, the Maxwell–Faraday equation links the temporal variations in the magnetic field to the spatial variations in the electric field (this is at the root of the phenomenon of electromagnetic induction), by means of equation [2.1c], which is here reproduced:

$$\vec{\nabla} \wedge \vec{E} = -\frac{\partial \vec{B}}{\partial t}$$

From this, we deduce that:

$$\vec{\nabla} \wedge \left(\vec{E} + \frac{\partial \vec{A}}{\partial t} \right) = \vec{0} \qquad [2.26]$$

Hence, vector analysis indicates that the electric field can be expressed in the form of the sum of the opposite of the temporal derivative of the vector potential and a zero-value rotational term, which can be expressed in the form of a gradient of a property which, in this context, is called the electric potential V. As $\vec{E} = -\vec{\nabla}.V$, generally speaking, we can write:

$$\vec{E} = -\vec{\nabla}.V - \frac{\partial \vec{A}}{\partial t} \qquad [2.27]$$

The term $\frac{\partial \vec{A}}{\partial t}$ is homogeneous to an electric field. It is called the Neumann field or the electromotive field, and reflects the phenomena of induction in mathematical form. It expresses the coupling between the electric field and the magnetic field. It is extremely important in the phenomena of propagation, all the more so at high frequencies.

Note that the vector potential is not unique; it is defined to within a given gradient $\left(\vec{A}' = \vec{A} + \vec{\nabla}.f\right)$, as is the scalar potential $\left(V' = V + const.\right)$. We can use this freedom to our advantage, and choose the vector potential which is most suitable to solve each problem we approach.

Equations [2.25] and [2.27] are connected to one another by the so-called *Lorenz gauge*:

$$\vec{\nabla}.\vec{A} + \varepsilon.\mu.\frac{\partial V}{\partial t} = 0 \qquad [2.28]$$

so:

$$\vec{\nabla}\left(\vec{\nabla}.\vec{A}\right) + \varepsilon.\mu.\frac{\partial \vec{\nabla}.V}{\partial t} = 0$$

and hence:

$$\vec{E} = -\frac{\partial \vec{A}}{\partial t} + \frac{1}{\varepsilon.\mu}\vec{\nabla}\left(\vec{\nabla}.\vec{A}\right) \qquad [2.29]$$

In the same way that the EM field, in a vacuum, propagates at the velocity $v = \frac{1}{\sqrt{\varepsilon.\mu}}$, the vector potential and the scalar potential also propagate at velocity v.

2.3.1. *Propagation equations for the vector potential*

We begin by assuming that the medium is homogeneous, isotropic, non-magnetic and non-polarizable, with the dielectric constant ε_0 and magnetic constant μ_0, so that:

$$\vec{J}_{pol} = \vec{J}_{mag} = \vec{0}\,,\ \rho_{mag} = \rho_{pol} = 0$$

To simplify the form of the expression, we set $\rho_{induced} = \rho$, $\vec{J} = \vec{J}_{source}$.

Maxwell's equations are written as follows:

– Maxwell–Gauss equation:

$$\vec{\nabla}.\vec{E} = \frac{\rho}{\varepsilon_0} \qquad [2.30a]$$

– Maxwell–Ampère equation:

$$\vec{\nabla} \wedge \vec{B} = \mu_0.(\vec{J} + \varepsilon_0.\frac{\partial \vec{E}}{\partial t}) \qquad [2.30b]$$

or indeed, given that $\vec{B} = \mu_0.\vec{H}$:

$$\vec{\nabla} \wedge \vec{H} = \vec{J} + \varepsilon_0.\frac{\partial \vec{E}}{\partial t}) \qquad [2.30c]$$

– Maxwell–Faraday equation:

$$\vec{\nabla} \wedge \vec{E} = -\mu_0 \frac{\partial \vec{H}}{\partial t} \qquad [2.30d]$$

– Maxwell–Thomson equation (magnetic flux equation):

$$\vec{\nabla}.\vec{H} = 0 \qquad [2.30e]$$

The Maxwell–Ampère equation [2.30b] is rewritten as follows, in light of equation [2.25]:

$$\vec{\nabla} \wedge \left(\vec{\nabla} \wedge \vec{A}\right) = \mu_0 .\vec{J} + \frac{1}{c^2}\frac{\partial \vec{E}}{\partial t} \qquad [2.31]$$

where $\varepsilon_0 .\mu_0 = \frac{1}{c^2}$ is the celerity of light.

Remember that $\vec{J} = \vec{J}_{source}$ is the "source" current density imposed by the tester.

In addition, the field $\vec{E}$ is deduced from the potentials V and $\vec{A}$ by means of relation [2.27], meaning that we can reformulate equation [2.31] as:

$$\vec{\nabla} \wedge \left(\vec{\nabla} \wedge \vec{A}\right) = \mu_0 .\vec{J} - \frac{1}{c^2}\left[\frac{\partial^2 \vec{A}}{\partial t^2} + \vec{\nabla}.\left(\frac{\partial V}{\partial t}\right)\right] \qquad [2.32]$$

Remember the following mathematical property (see Appendix 2):

$$\Delta.\vec{A} = \vec{\nabla}.\left(\vec{\nabla}.\vec{A}\right) - \vec{\nabla} \wedge \left(\vec{\nabla} \wedge \vec{A}\right)$$

which is the Laplace function of the vector $\vec{A}$ with the following components:

$$\Delta.\vec{A} = \begin{cases} \Delta A_x = \left(\frac{\partial^2}{\partial x^2} + \frac{\partial^2}{\partial y^2} + \frac{\partial^2}{\partial z^2}\right) A_x \\ \Delta A_y = \left(\frac{\partial^2}{\partial x^2} + \frac{\partial^2}{\partial y^2} + \frac{\partial^2}{\partial z^2}\right) A_y \\ \Delta A_z = (\ldots\ldots\ldots\ldots\ldots\ldots\ldots\ldots) A_z \end{cases}$$

After reorganizing the terms, equation [2.32] can be written as:

$$\Delta\vec{A} - \frac{1}{c^2}\frac{\partial^2 \vec{A}}{\partial t^2} = -\mu_0 .\vec{J} + \vec{\nabla}\left(\vec{\nabla}.\vec{A} + \frac{1}{c^2}\frac{\partial V}{\partial t}\right) \qquad [2.33]$$

This equation is complicated, and still simultaneously involves $\vec{A}$ and V. Thus, we can simplify it by making the following consideration: $\vec{A}$ must be such that its rotational is equal to the magnetic induction field $\vec{B}$. However, the vector potential $\vec{A}$ is defined to within a gradient. Hence, we can exploit this freedom in the definition of $\vec{A}$ to *choose* its definition, using the Lorenz gauge:

$$\vec{\nabla}.\vec{A}+\frac{1}{c^2}\frac{\partial V}{\partial t}=0$$

The partial differential equation satisfied by the vector potential $\vec{A}$ is then simply:

$$\Delta.\vec{A}-\frac{1}{c^2}\frac{\partial^2\vec{A}}{\partial t^2}=-\mu_0.\vec{J} \qquad [2.34]$$

The form of this equation is very elegant. Firstly, the differentials with respect to x, y, z and t play a symmetrical role. Secondly, in the case of currents that are constant over time, $\vec{A}$ itself is constant, and equation [2.34] is identical to that which we would obtain in magnetostatics:

$$\Delta.\vec{A}=-\mu_0.\vec{J} \qquad [2.35]$$

2.3.2. *Propagation equations for the scalar potential*

The Maxwell–Gauss equation [2.30a] is expressed as a function of the potentials, in light of equation [2.27]:

$$\vec{\nabla}.\left(\frac{\partial\vec{A}}{\partial t}+\vec{\nabla}.V\right)=-\frac{\rho}{\varepsilon_0} \qquad [2.36]$$

The gauge equation [2.28] is differentiated with respect to time, and assumes the form:

$$\vec{\nabla}.\left(\frac{\partial\vec{A}}{\partial t}\right)=-\frac{1}{c^2}\frac{\partial^2 V}{\partial t^2} \qquad [2.37]$$

Using the property of the Laplacian, which is valid for any scalar field:

$$\Delta.V=\vec{\nabla}.\left(\vec{\nabla}.V\right) \qquad [2.38]$$

we obtain the propagation equation with a source for the scalar potential:

$$\Delta.V - \frac{1}{c^2}\frac{\partial^2 V}{\partial t^2} = -\frac{\rho}{\varepsilon_0} \quad [2.39]$$

The potentials conform to propagation equations of the type:

$$\Delta.f - \frac{1}{c^2}\frac{\partial^2 f}{\partial t^2} = S \quad [2.40]$$

where S is the source term, and c the celerity of light.

Equations [2.34] and [2.39] are used as the basis for the development of computer codes for antennas.

2.3.3. *Vector and scalar potentials in the harmonic regime*[8]

The variation of the fields is in $e^{j.\omega.t}$. It follows that $\frac{\partial^2}{\partial t^2} \Rightarrow -\omega^2$, and equations [2.34] and [2.39] are written as:

$$\Delta.\vec{A} + k^2.\vec{A} = -\mu_0.\vec{J} \quad [2.41a]$$

$$\Delta.V + k^2.V = -\frac{\rho}{\varepsilon_0} \quad [2.41b]$$

Equations [2.41a] and [2.41b] are the wave equations, or Helmholtz equations.

They demonstrate that we can attach the vector potential to the currents and the scalar potential to the charges. It can also be shown that the fields $\vec{E}$ and $\vec{H}$ also satisfy a similar wave equation, although the independent terms are rather more complicated.

In a vacuum, fields propagate at the velocity $c = \frac{1}{\sqrt{\varepsilon_0.\mu_0}}$ = 299,800 km/s, where

$k^2 = \omega^2.\varepsilon_0.\mu_0 = \frac{\omega^2}{c^2} = \left(\frac{2.\pi}{\lambda}\right)^2$ k is called the propagation constant, $v = \lambda.f_{rq}$ with

8 Silver 1984.

$v = \dfrac{1}{\sqrt{\varepsilon.\mu}} < c$, v is the wave velocity in the medium with constants ε and μ, with λ being the wavelength at frequency f_{rq}, and ω the pulsation (in radians/second) $\omega = 2\pi/f_{rq}$.

2.4. Propagation equation for fields $\vec{E}$ and $\vec{H}$ [9]

Generally, in an LHI medium with permittivity ε and permeability μ:

$$\Delta.\vec{E} - \frac{1}{v^2}\frac{\partial^2 \vec{E}}{\partial t^2} = \mu.\frac{\partial \vec{J}}{\partial t} + \vec{\nabla} \wedge \vec{J}_{mag} + \frac{1}{\varepsilon}\vec{\nabla}.\rho \qquad [2.42a]$$

$$\Delta.\vec{H} - \frac{1}{v^2}\frac{\partial^2 \vec{H}}{\partial t^2} = \mu.\frac{\partial \vec{J}_{mag}}{\partial t} - \vec{\nabla} \wedge \vec{J} + \frac{1}{\mu}\vec{\nabla}.\rho_{mag} \qquad [2.42b]$$

In a medium without a source of current and charge:

$$\Delta.\vec{E} - \frac{1}{v^2}\frac{\partial^2 \vec{E}}{\partial t^2} = \vec{0} \qquad [2.43a]$$

$$\Delta.\vec{H} - \frac{1}{v^2}\frac{\partial^2 \vec{H}}{\partial t^2} = \vec{0} \qquad [2.43b]$$

Where the fields vary over time like $e^{j.\omega.t}$, then $\dfrac{\partial^2}{\partial t^2}$ is replaced by $-\omega^2$. Each component of $\vec{E}$ and $\vec{H}$ satisfies the Helmholtz scalar equation, expressed in the following form:

$$\Delta.\Psi + k^2.\Psi = -\delta(x).\delta(y).\delta(z) \qquad [2.44]$$

When the point source is a Dirac pulse $\delta(x,y,z)$, owing to spherical symmetry, the Laplacian is written in spherical coordinates, and Ψ is solely dependent on the radial data r in the direction of wave propagation.

9 Silver 1984.

Two solutions to equation [2.44] are:

$$\Psi = \frac{1}{4.\pi.r} e^{-j.k.r} \text{ and } \Psi = \frac{1}{4.\pi.r} e^{+j.k.r}$$

which correspond, respectively, to waves propagating radially *forwards* and *backwards* from the source point, which is the origin of the coordinate axes $< o, x, y, z >$ shown in Figure 2.1.

2.5. Solving the Helmholtz equations for the vector and scalar potentials

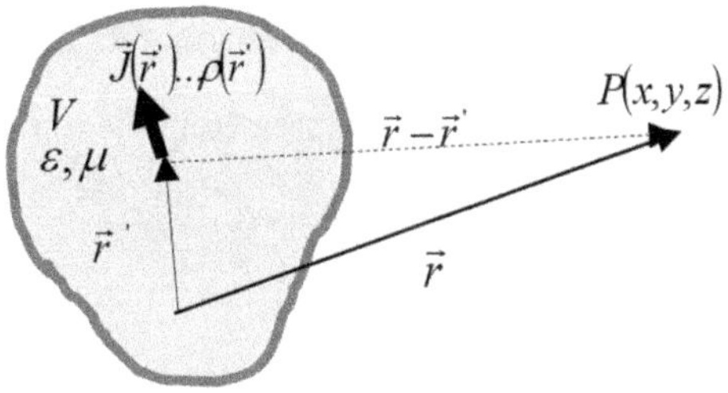

Figure 2.3. *Radiation of a distribution of charges and currents*

In the coordinate system shown in Figure 2.1, we assume we have sources limited to a volume V surrounded by an infinite homogeneous and lossless medium ($\varepsilon = \varepsilon_0$, $\mu = \mu_0$, $\sigma = 0$). These sources of currents and charges in the volume V are those which are distributed across the conductive surfaces and/or in the dialectics of which the antenna we are studying is built. The potentials at a point $\vec{r}$ are obtained by integration of the Helmholtz equations [2.41a] and [2.41b] in spherical coordinates:

$$\vec{A}(\vec{r}) = \frac{\mu}{4.\pi} \iiint_V \vec{J}(\vec{r}') \frac{e^{-j.k.\left|\vec{r}-\vec{r}'\right|}}{\left|\vec{r}-\vec{r}'\right|}.dv' \qquad [2.45a]$$

$$V(\vec{r}) = \frac{1}{4.\pi.\varepsilon} \iiint_V \rho(\vec{r}') \frac{e^{-j.k.\left|\vec{r}-\vec{r}'\right|}}{\left|\vec{r}-\vec{r}'\right|}.dv' \qquad [2.45b]$$

It can be said that the potentials are given by superposition of spherical waves (decrease in $\frac{1}{r}$) having a propagation constant $k = \omega.\sqrt{\varepsilon.\mu}$ and a wavelength $\lambda = \frac{2.\pi}{k}$. In the above expressions, the radial vector $\vec{r}$ represents any point, which may be far or near and even inside the volume containing the sources. Once we know the potentials, we can immediately calculate the fields $\vec{E}$ and $\vec{H}$ using equations [2.25] and [2.29], recapped here:

$$\vec{B} = \vec{\nabla} \wedge \vec{A} \text{ and } \vec{E} = -\frac{\partial \vec{A}}{\partial t} + \frac{1}{\varepsilon.\mu}\vec{\nabla}\left(\vec{\nabla}.\vec{A}\right)$$

If we set $R = \left|\vec{R}\right| = \left|\vec{r} - \vec{r}'\right|, \hat{u} = \frac{\vec{R}}{R}$, in the harmonic regime, to simplify the mathematical formulation, the sources are expressed in the form:

$$\vec{J}\left(\vec{r}',t\right) = \vec{J}\left(\vec{r}'\right)e^{j\omega\left(t - \frac{R}{v}\right)} = \vec{J}\left(\vec{r}'\right)e^{j(\omega t - kR)} \qquad [2.46]$$

For the field $\vec{E}$, we obtain:

$$\vec{E}\left(\vec{r}\right) = -\frac{j.\omega.\mu}{4.\pi}.\iiint_V \vec{J}\left(\vec{r}'\right).\left[1 + \frac{\vec{\nabla}.\vec{\nabla}}{k^2}\right].\frac{e^{-j.k.R}}{R}.dv \qquad [2.47a]$$

This equation is known as the electric-field integral equation (EFIE). It is a fundamental equation used in the calculation and design of antennas. With it, we can calculate the radiation of an antenna when the current distribution is known. Nevertheless, we do still need to calculate this current distribution, and therein lies the difficulty. Generally speaking, it is not possible to find a solution analytically, except for simple structures used as references (Silver 1984; Milligan 2005; Huang 2008; Balanis 2008), such as fine dipoles and other wired variants. In more complicated situations, the solution generally requires the use of relatively complex numerical codes, such as the method of moments (MoM), the finite element method (FEM), the finite-difference time-domain (FDTD) method, etc.

When the geometries at hand are extremely complex, advanced numerical methods must be used. One proven method is based on the equivalence principle (Figures 1.14a and 1.14b). The elements of which the antenna is to be built are sampled, including the volume V delimited by the surface Σ that surrounds it. The surface Σ surrounding the volume V involves tangential components of the EM field, as well as the equivalent electric and magnetic currents (fictitious). There is a wide range of formulations, but most of them apply only to specific problems. The EFIE equation is preferable, because it is stable in the case of planar structures or structures whose radii of curvature are much greater than the wavelength. However, it is less reliable when we are dealing with very low frequencies. The EFIE formulation is not apt for use with enclosed PEC surfaces, unlike the MFIE formulation [2.47b].

For the field $\vec{H}$, if we set:

$$\psi(\vec{r}-\vec{r}\,')=\frac{e^{-\vec{k}.|\vec{r}-\vec{r}\,'|}}{|\vec{r}-\vec{r}\,'|}=\psi(\vec{R})=\frac{e^{-jk.R}}{R}$$

and:

$$\psi_1(\vec{R})=-\left(j.k+\frac{1}{R}\right)\psi(\vec{R})$$

and use:

$$\vec{\nabla}\wedge\left[\psi(\vec{R}).\vec{J}(\vec{r}\,')\right].=\psi_1(\vec{R}).\hat{u}\wedge\vec{J}(\vec{r}\,')$$

then the MFIE equation is written thus:

$$\vec{H}(\vec{r})=\frac{j.k}{4.\pi}\iiint_V\left(1+\frac{1}{j.k.R}\right)\left(\vec{J}(\vec{r}\,').\wedge\hat{u}\right).\psi(\vec{R})\,dv \qquad [2.47b]$$

Equations [2.47a] and [2.47b] apply to the currents contained with the volume V. They can be used to calculate the EM field throughout the space, include fields nearby to Σ, by use of the equivalence theorem.

2.5.1. *Orthogonality of distance fields zone and radiated power; radiation pattern*[10]

For the time being, we shall focus our attention on observation points that are very far removed from the antenna (in comparison to *its linear dimensions*). This relative distance is referred to in the existing literature as a *far field*. The distance

10 Stutzman 1981; Silver 1984.

between source and observer $R=\left|\vec{r}-\vec{r}^{'}\right|$ in equations [2.47a] and [2.47b] plays two different roles: as the *amplitude* in the denominator, and as the *phase* within the exponent. If $\left|\vec{r}^{'}\right| << \left|\vec{r}\right|$, we can replace the exact expression of this distance with a limited development. For the amplitude, the dominant term $\left|\vec{r}-\vec{r}^{'}\right| \approx \left|\vec{r}\right|$ is sufficient. On the other hand, the phase term must be more accurate, with an additional term, making the following approximation in the exponent. In the coordinate system shown in Figure 2.1:

$$\left|\vec{r}-\vec{r}^{'}\right| \approx r-\widehat{e}_r.\vec{r}^{'} = r-x^{'}\sin(\theta).\cos(\varphi)-y^{'}\sin(\theta).\sin(\varphi)-z^{'}\cos(\theta)$$

where $\widehat{e}_r$ is the unit vector of $\vec{r}$ ($\vec{r} = \widehat{e}_r.r$).

These approximate relations are obvious from a geometric standpoint, when we consider that the vectors $\vec{r}$ and $(\vec{r}-\vec{r}^{'})$ are practically parallel. These approximations introduce a major simplification in the calculation of $\vec{A}$, and therefore of the fields $\vec{E}$ and $\vec{H}$:

$$\vec{E}(\vec{r}) = -\frac{j.k^2}{4.\pi.\omega.\varepsilon}.\iiint_V \left[(\vec{J}(\vec{r}^{'})\wedge\widehat{u})\wedge\widehat{u}\right].\frac{e^{-j.k.r}}{r}dv \qquad [2.48a]$$

$$\vec{H}(\vec{r}) = -\frac{j.k}{4.\pi}\iiint_V \left(\vec{J}(\vec{r}^{'})\wedge\widehat{u}\right).\frac{e^{-j.k.r}}{r}dv \qquad [2.48b]$$

We can see that the two fields $\vec{E}(\vec{r})$ and $\vec{H}(\vec{r})$ are orthogonal and connected by the relation:

$$\vec{H}(\vec{r}) = \frac{\widehat{u}\wedge\vec{E}(\vec{r})}{\eta} \qquad [2.49]$$

where $\eta = \sqrt{\frac{\mu}{\varepsilon}}$ is the *characteristic* impedance of the medium. In a vacuum, $\eta \approx 377.\Omega$:

$$\widehat{u} = \frac{\vec{R}}{\left|\vec{R}\right|}$$

When we set:

$$\vec{E}(\vec{r}) = -\frac{j.k^2}{4.\pi.\omega.\varepsilon}\frac{e^{-j.k.r}}{r}\vec{f}(\theta,\varphi) \quad [2.50a]$$

$$\vec{f}(\theta,\varphi) = \iiint_V \left[(\vec{J}(\vec{r}') \wedge \hat{u}) \wedge \hat{u}\right] dv \quad [2.50b]$$

the vector integral (also known as the *radiation integral*) $\vec{f}(\theta,\varphi)$ gives the angular dependence of the electric field. It plays a crucial role in the theory underlying antennas. This integral has no analytical solution except in the simple reference cases mentioned above.

It should be noted, in passing, that in a medium with loss – i.e. where its conductivity is non-negligible and/or its permittivity is complex – the characteristic impedance is complex and is given by Huang (2008):

$$\eta = \frac{|\vec{E}|}{|\vec{H}|} = \sqrt{\frac{\omega.\mu}{\omega.\varepsilon - j.\sigma}} \quad [2.51]$$

The wavenumber k is complex:

$$k = \sqrt{j.\omega.\mu.(\sigma + j.\omega\varepsilon)} = \alpha + j.\beta \quad [2.52]$$

$$\alpha = \omega.\sqrt{\mu.\varepsilon}.\left[\frac{1}{2}.(\sqrt{1+\frac{\sigma^2}{\varepsilon^2.\omega^2}} - 1)\right]^{\frac{1}{2}}$$

$$\beta = \omega.\sqrt{\mu.\varepsilon}.\left[\frac{1}{2}.(\sqrt{1+\frac{\sigma^2}{\varepsilon^2.\omega^2}} + 1)\right]^{\frac{1}{2}}$$

We can see that in the far zone, there is no radial component for the fields. The Poynting vector is given by:

$$\vec{S}(\vec{r}) = \vec{E}(\vec{r}) \wedge \vec{H}(\vec{r}) = \frac{1}{\eta}|\vec{E}|^2.\hat{e}_r \quad [2.53]$$

where $\hat{e}_r = \frac{\vec{r}}{r}$ is the unit vector of $\vec{r}$ from the origin of the axes at the observation point P.

The Poynting vector, in an isotropic medium, indicates the direction of propagation of an electromagnetic wave. The flux of the Poynting vector across a surface, closed or otherwise, is equal to the power carried by the wave across that surface. The modulus of this vector, therefore, is a power per unit surface – i.e. an energy flux – and is expressed in watts per square meter.

Because the Poynting vector is the vector product of two fields, it is orthogonal to these two fields. This triplet represents a *right-handed* coordinate system $(\vec{E}, \vec{H}, \vec{S})$.

This radial component gives us the *radiated power density* in the direction of the wave propagation, which can be written in the form:

$$W(r,\theta,\varphi) = \vec{S}.\hat{e}_r = \frac{1}{\eta}\left(\left|\vec{E}_\theta\right|^2 + \left|\vec{E}_\varphi\right|^2\right) \text{ (W/m}^2\text{)} \qquad [2.54]$$

where $\vec{f}(\theta,\varphi) = \vec{f}_\theta.\hat{\theta} + \vec{f}_\varphi.\hat{\varphi}$:

$$W(r,\theta,\varphi) = \frac{\eta}{4.\lambda^2.r^2}\left(\left|f_\theta\right|^2 + \left|f_\varphi\right|^2\right) \text{(W/m}^2\text{)} \qquad [2.55]$$

Much like the fields, the power density depends on the three spherical coordinates, but the dependence on the radial coordinate is always the same (inversely proportional to the square of the distance). We then define the intensity of radiation U as:

$$U(\theta,\varphi) = r^2.W(r,\theta,\varphi) = \frac{\eta}{4.\lambda^2}\left(\left|f_\theta\right|^2 + \left|f_\varphi\right|^2\right) \text{(W/steradian)} \qquad [2.56]$$

There is always a particular direction (θ_0, φ_0) in which the radiation intensity is greatest:

$$U_{MAX} = U(\theta_0, \varphi_0)$$

The unit-normed radiation diagram $r(\theta,\varphi)$ is defined by the dimensionless ratio:

$$r(\theta,\varphi)=\frac{U(\theta,\varphi)}{U_{MAX}} \qquad [2.57]$$

2.6. Harmonic form of Maxwell's equations[11]

With LHI and ohmic materials, the study of antennas is conducted in almost all cases with a variation of the fields in $e^{j.\omega.t}$. This being the case, Maxwell's equations are written as follows:

$$\vec{\nabla}\wedge\vec{E}+j.\omega\mu.\vec{H}=-\vec{J}_{mag} \qquad [2.58a]$$

$$\vec{\nabla}\wedge\vec{H}=(\sigma+j.\omega.\varepsilon)\vec{E}+\vec{J}_{source} \qquad [2.58b]$$

$$\vec{\nabla}.(\mu.\vec{H})=\rho_{mag} \qquad [2.58c]$$

$$\vec{\nabla}.(\varepsilon.\vec{E})=\rho \qquad [2.58d]$$

$$\vec{\nabla}.\vec{J}+j\omega.\rho=0 \qquad [2.58e]$$

$$\vec{\nabla}.\vec{J}_{mag}+j\omega.\rho_{mag}=0 \qquad [2.58f]$$

Note that for an inhomogeneous medium, ε and μ are functions of the position in the volume in equations [2.58c] and [2.58d]. It should be noted that the continuity equation [2.58e] determines the charge density directly on the basis of the current density.

2.7. Physical interpretation of the Poynting theorem[12]

2.7.1. *Poynting vector in the time domain*

The propagation medium is an LHI medium with electric permittivity ε and magnetic permeability μ. On the basis of equations [2.1b] and [2.1c], recapped below with the following conditions, to simplify the formulae:

11 Silver 1984; Huang 2008.

12 Silver 1984; Huang 2008.

$$\vec{E}(x,y,z,t)=\vec{E}$$

$$\vec{H}(x,y,z,t)=\vec{H}$$

$$\vec{\nabla}\wedge\vec{H}=\vec{J}+\frac{\partial\vec{D}}{\partial t}$$

$$\vec{\nabla}\wedge\vec{E}=-\frac{\partial\vec{B}}{\partial t}$$

when we set:

$$\vec{H}.\left(\vec{\nabla}\wedge\vec{E}\right)-\vec{E}.\left(\vec{\nabla}\wedge\vec{H}\right)=-\vec{E}.\vec{J}=-\frac{\partial}{\partial t}\left(\frac{\vec{E}.\vec{D}}{2}+\frac{\vec{H}.\vec{B}}{2}\right)$$

and using the relations:

$$\varepsilon=\frac{\left|\vec{D}\right|}{\left|\vec{E}\right|}=\frac{D}{E}\text{ and }\mu=\frac{\left|\vec{B}\right|}{\left|\vec{H}\right|}=\frac{B}{H}$$

so:

$$\vec{H}.\left(\vec{\nabla}\wedge\vec{E}\right)-\vec{E}.\left(\vec{\nabla}\wedge\vec{H}\right)=-\vec{E}.\vec{J}=-\frac{\partial}{\partial t}\left(\varepsilon.\frac{E^2}{2}+\mu.\frac{H^2}{2}\right) \qquad [2.59]$$

which is the expression of the Poynting theorem in local form for a volume dv with $\partial W=\left(\varepsilon.\frac{E^2}{2}+\mu.\frac{H^2}{2}\right)$:

$$-\vec{\nabla}.\left(\vec{E}\wedge\vec{H}\right)-\vec{E}.\vec{J}=\frac{\partial W}{\partial t}=\frac{\partial}{\partial t}\left(\varepsilon.\frac{E^2}{2}+\mu.\frac{H^2}{2}\right) \qquad [2.60]$$

Formally, this equation resembles the continuity equation [2.58e]. It expresses conservation of energy rather than of charge. The Poynting vector is expressed by formula [2.53]. The values $\varepsilon\frac{E^2}{2}$ and $\mu\frac{H^2}{2}$ represent the electrical and magnetic energy densities, respectively. The term $\vec{E}.\vec{J}$ measures the portion of electromagnetic energy dissipated or produced per unit volume. If $\vec{E}.\vec{J}$ is positive, the energy is dissipated; if it is negative, it represents the production of electromagnetic energy.

The local equation [2.60], integrated into a volume V delimited by the surface Σ, is expressed by:

$$-\iiint_V \frac{\partial W}{\partial t} dv = \iiint_V \vec{\nabla}.\left(\vec{E} \wedge \vec{H}\right) dv + \iiint_V \vec{E}.\vec{J} dv \qquad [2.61]$$

And, by means of application of the Ostrogradski theorem, equation [2.61] enables us to give a physical interpretation to the Poynting vector:

$$-\iint_\Sigma (\vec{E} \wedge \vec{H}).\vec{ds} = \frac{\partial W}{\partial t} + \iiint_V \vec{E}.\vec{J}.dv$$

which is written:

$$-\iint_\Sigma \vec{S}.\vec{ds} = \frac{\partial W}{\partial t} + \iiint_V \vec{E}.\vec{J}.dv \qquad [2.62]$$

where $\vec{ds} = ds.\hat{n}$ is the surface element ds and $\hat{n}$ is the normal orientated toward the outside of the volume V.

The term on the left-hand side represents the *incoming* flux of the Poynting vector, in light of the negative sign. The first term on the right-hand sign is the variation of the energy on the volume V. The second term represents the power supplied to the currents within the volume V. This may be linked to the Joule effect or to the acceleration of the moving charged particles. Equation [2.62] expresses the energy equilibrium of the volume V. The radiated power entering into the volume V is equal to the sum of the increase in energy of V per unit time and the power applied to the particles within it. Thus, the flux of the Poynting vector is interpreted as the power radiated across a surface, by the following expression:

$$P(t) = \iint_\Sigma \vec{S}.\vec{ds} \qquad [2.63]$$

For the study of antennas (Picon 2009), this equation written in a different order leads us to a helpful interpretation:

$$-\frac{\partial W(t)}{\partial t}=\iint_{\Sigma}\vec{S}.d\vec{s}+\iiint_{V}\vec{E}.\vec{J}.dv$$

Indeed, the loss of EM power in V is equal to the radiated power exiting the volume V, across Σ, added to the power transferred by the Joule effect into V. The local equation of the Poynting vector is expressed by:

$$\vec{\nabla}.\vec{S}(t)+\frac{\partial W(t)}{\partial t}=-\vec{E}(t).\vec{J}(t) \qquad [2.64]$$

which is the *local* conservation of energy equation, according to the interpretation above.

2.7.2. *Poynting vector in the frequency domain*[13]

In the frequency domain, fields are considered at a given frequency. They vary in a sinusoidal fashion. We use the complex representation:

$$\vec{E}(x,y,z,t)=\vec{E}(t)=\left(\vec{E}_{réel}+j\vec{E}_{im}\right)e^{j.\omega.t} \qquad [2.65a]$$

$$\vec{H}(x,y,z,t)=\vec{H}(t)=\left(\vec{H}_{réel}+j\vec{H}_{im}\right)e^{j.\omega.t} \qquad [2.65b]$$

It is the average power which is of interest, rather than the instantaneous power:

$$\vec{S}_{average}=\frac{1}{\tau}\int_{0}^{\tau}\vec{S}dt$$

where $\left(\tau=\frac{2.\pi}{\omega}\right)$ is the period.

13 Silver 1984; Picon 2009.

Thus:

$$\vec{S}_{average} = \frac{1}{2}\left(\vec{E}_{real} \wedge \vec{H}_{real} + \vec{E}_{im} \wedge \vec{H}_{im}\right)$$

$$\vec{S}_{average} = \frac{1}{2}\Re eal\left(\vec{E} \wedge \vec{H}^{*}\right) \text{ in } \left(\text{w/m}^2\right)$$

The power across a surface Σ is:

$$\vec{S}_{average} = \frac{1}{2}\Re eal.\iint_{\Sigma}\left(\vec{E} \wedge \vec{H}^{*}\right)d\vec{s} \qquad [2.66]$$

where $\vec{H}^{*}$ is the complex conjugate of $\vec{H}$.

2.8. Polarized wave[14]

2.8.1. *Definition of a plane wave*

The relation between the electric field and the magnetic field which satisfies equation [2.49], recapped below, indicates that at a significant distance from the sources, the EM wave propagates in the direction $\hat{e}_r$ of the Poynting vector $\vec{S}$. This wave has a constant phase front that evolves radially in $e^{-j.\omega.t}.e^{j.k.r}$. A wave such as this is said to be *plane*. A plane wave, theoretically, propagates without attenuation, but this does not reflect the laws of physics. Indeed, on the basis of Maxwell's equations, the radiation condition (Statton 1941; Silver 1984) imposes a null field infinitely. More specifically, at a great distance, the wave is said to be *locally plane*, in view of the presence of the spherical term $1/r$, which ensures that any field is zero at infinity:

$$\vec{H} = \frac{1}{\eta}\hat{e}_r \wedge \vec{E}$$

A condition for a wave to be plane is that the radial component of the EM field at the observation point is null, meaning that it is simultaneously TE and TM, so TEM.

14 Milligan 2005.

The expression of the electric field is given by:

$$\vec{E}(\vec{r},t) = \vec{E}_0.e^{j.(\omega.t - k.R)} \quad [2.67]$$

where $\vec{k}$ is the wavenumber and $\vec{r} = r.\hat{e}_r$ the position vector of the observation point (x, y, z) in relation to the origin of the axes.

On the sphere whose center o is the origin of the axes, along $\vec{r}_r$, $\vec{E}_0$ is independent of both position and time t. The surface of the sphere is a surface of equal phase called the *wavefront* or *wave surface*. $\vec{H}(\vec{r},t)$ is a solution to equation [2.48b], and satisfies Maxwell's equation [2.58a], where $\vec{J}_{mag} = \vec{0}$, which echoes the formula [2.53].

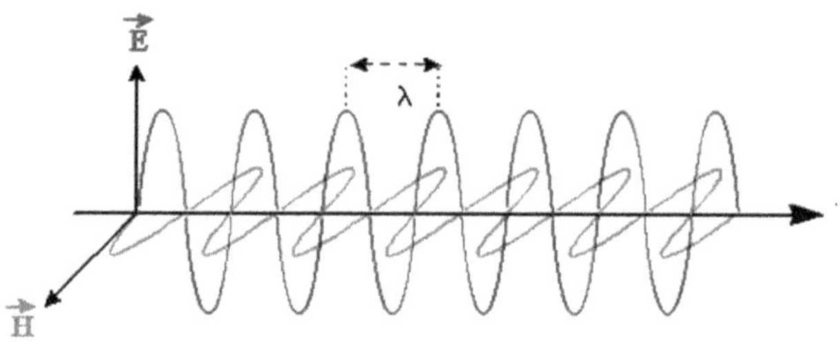

Figure 2.4. *Propagation of a linearly polarized plane wave*

2.8.2. *Polarizations of a wave*

Equations [2.47a] and [2.47b], where the dependence on time t is omitted, can be used to instantly calculate the electric field $\vec{E}(\vec{r})$ and magnetic field $\vec{H}(\vec{r})$ at a point $P(\vec{r})$ in space. The vector radiation pattern depends solely on the distribution of the current "density" vectors $\vec{J}$ across the radiating structure contained within the volume V. We saw earlier that, with the LHI materials used to build an antenna, the superposition theorem applies, meaning that we add together the current densities across the integral. We must therefore expect that at any given time t, the vector polarization of the radiated EM field will depend on the sum of the orientation of those vectors $\vec{J}$. The elementary current density $\vec{J}(\vec{r},t)$ at each point on the radiating surface of the antenna is a vector that can be broken down into any two vectors such that $\vec{J}(\vec{r},t)=\vec{J}_1(\vec{r},t)+\vec{J}_2(\vec{r},t)$. Equations [2.48a] and [2.48b] tell us the components of the EM field with these two distributions. Clearly, if all these elementary currents are constant and have the same orientation, $\vec{J}(\vec{r},t)$ can be removed from the integral, and represents all of the currents on the structure, giving us $\vec{E}(r,\theta,\varphi,t)$ and $\vec{H}(r,\theta,\varphi,t)$, an orientation which varies over time and is precisely defined by (r,θ,φ).

What is of interest to telecommunications operators, who are the main users of antennas, particularly for long-distance services that are regulated by the International Telecommunications Union (ITU), is to increase the possibility of using the same frequency on two orthogonal polarizations in order to increase the capacity of their networks and thus gain a greater return on their investments. This is primarily the case for satellite services, using terrestrial radio beams. With regard to base transceiver stations (BTSs) for mobiles, given the presence of multiple obstacles along the path the signal needs to travel, one or two linear polarizations orientated at ±45° from the vertical from the ground are needed. With applications for space exploration, the strategy is known as *frequency reuse* by polarization discrimination. For a given frequency, the use of two orthogonal polarizations doubles the capacity of a network. For a circular polarization in the C band, for example (see Table 1.2), the standard imposed by the ITU is –30.7 dB, which is a ratio of ellipticity of $RE = 1.06$. For a linear polarization in the K_u band, for example (see Table 1.2), the ratio between the nominal polarization and the cross-polarization is generally > 35 dB, always within the angular cone of satellite tracking accuracy. It is clear that this type of specification cannot be preserved with miniaturized antennas on whose surface the currents are not strictly ordered in specific directions.

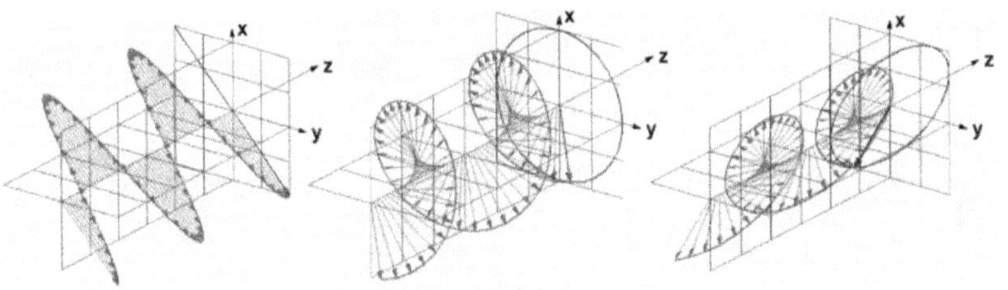

a) b) c)

Figure 2.5. *Propagation of a) linear polarization; b) circular polarization; c) elliptical polarization*

At a time t, the polarization of a wave is given by the orientation of the electric field $\vec{E}$ across a sphere with radius r in the far field zone. On that sphere, $\vec{E}$ is dependent only on the angles θ and φ, so:

$$\vec{E}(\theta,\varphi,t) = E_\theta(t).\hat{\theta} + E_\varphi(t).\hat{\varphi}$$

E_θ and E_φ are the components in the directions of the unit vectors $\hat{\theta}$ and $\hat{\varphi}$. $\vec{E}$ is expressed by:

$$\vec{E}(\theta,\varphi,t) = \vec{E}_\theta(t) + \vec{E}_\varphi(t) = E_1 e^{-j\left(\omega.t+\frac{2\pi}{\lambda}.r\right)}.\hat{\theta} + \delta.E_2 e^{-j\left(\omega.t+\frac{2\pi}{\lambda}.r\right)}.\hat{\varphi} \qquad [2.68]$$

δ is a complex number on which the type of polarization depends: it may be linear, elliptical, left circular or right circular. In the normalized terminology of the ITU, this is LHCP (left-hand circular polarization) or RHCP (right-hand circular polarization).

E_1 and E_2 are the real amplitudes of the complex components E_θ and E_φ.

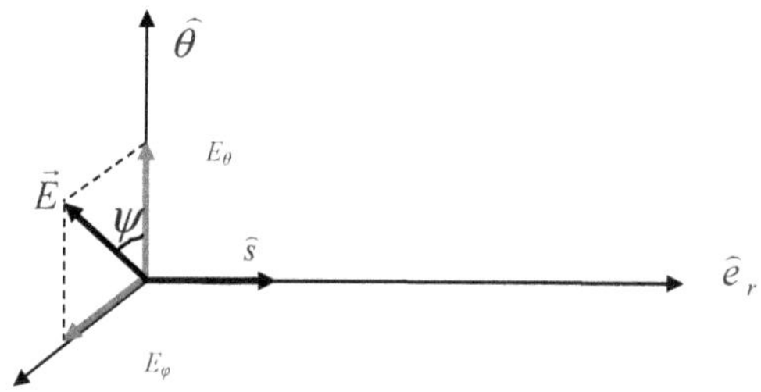

Figure 2.6. *Representation of the vector electric field*

To simplify the formulation, we can return to polar coordinates to express $\vec{E}$:

$$\vec{E} = E_0\left[\cos(\psi).\hat{\theta} + \delta\sin(\psi).\hat{\varphi}\right]e^{-j\left(\omega.t+\frac{2\pi}{\lambda}r\right)}$$

so, if we omit the exponent:

$$\vec{E} = E_0\left[\cos(\psi).\hat{\theta} + \delta\sin(\psi).\hat{\varphi}\right] \qquad [2.69]$$

where E_0 is the real amplitude.

The polarization is linear if δ is real. In this case, the wave propagates along the $\hat{e}_r$ axis, which is the direction of propagation, with the field $\vec{E}$ in the plane with inclination ψ.

Whilst the theory remains generally applicable, let us begin by supposing that $\vec{E}$ is such that $\psi = 0$:

$$\vec{E} = \vec{E}_\theta = E_0.\hat{\theta} \qquad [2.70]$$

If the plane $< \hat{\theta}, \hat{e}_r >$ is vertical in relation to the ground, sticking with the ITU's definition, then the polarization is said to be *vertical*.

If $\vec{E}$ is such that $\psi = \pi/2$ with $\delta = 1$ (to simplify the formula):

$$\vec{E} = \vec{E}_\varphi = E_0.\hat{\varphi} \qquad [2.71]$$

In this case, the polarization is said to be *horizontal*.

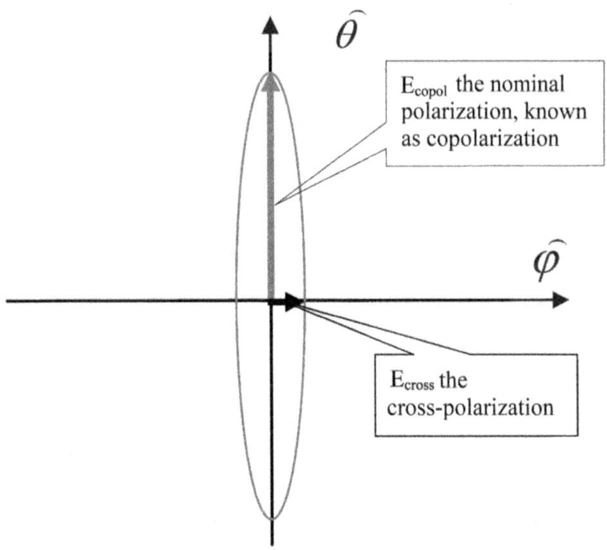

Figure 2.7. *Imperfect linear polarization*

In fact, in reality, polarization is never purely linear or purely circular, as the current densities $\vec{J}_1(\vec{r},t)$ and $\vec{J}_2(\vec{r},t)$ do not have constant orientation and amplitude, due to imperfections in the manufacture of the antenna or the theoretical impossibility in the case of antennas smaller than the wavelength λ. With respect to the problem of feasibility of an antenna to obtain effective polarization discrimination, we must take account of the surrounding medium through which the EM wave must pass: the effect of hydrometeors, the upper atmospheric layers, the presence of obstacles, etc. Consequently, the field $\vec{E}$ is always slightly elliptical, rotating randomly to the left or right for polarization said to be linear. δ is therefore complex, such that its modulus $|\delta| << 1$.

The polarization decoupling in the case of linear polarization is given by:

$$D = 20Log_{10}\left(\frac{E_{cross}}{E_{copol}}\right) \qquad [2.72]$$

1) *The polarization is elliptical*: in this case, it is the sum of two linear polarizations with a phase shift between them of $\pm\frac{\pi}{2}$ in the direction of RHCP or LHCP.

2) *The polarization is purely circular if* $\delta = \pm j$: in the axis system $<\hat{\theta},\hat{\varphi}>$ or $<\hat{x},\hat{y}>$, purely circular polarization is expressed by:

$$\hat{R} = RHCP = \frac{1}{2}\left(\hat{\theta} - j\hat{\varphi}\right) \text{ or } \hat{R} = RHCP = \frac{1}{2}\left(\hat{x} - j.\hat{y}\right) \qquad [2.73a]$$

$$\hat{L} = LHCP = \frac{1}{2}\left(\hat{\theta} + j\hat{\varphi}\right) \text{ or } \hat{L} = LHCP = \frac{1}{2}\left(\hat{x} + j.\hat{y}\right) \qquad [2.73b]$$

$$\vec{E} = E_L.\hat{L} + E_R.\hat{R} \qquad [2.74]$$

Let us set $E_L = \vec{E}.\hat{L}^*$ and $E_R = \vec{E}.\hat{R}^*$, where $\hat{R}^*$ is the conjugate vector of $\hat{R}$, and $\hat{L}^*$ is the conjugate vector of $\hat{L}$.

We can verify that $\widehat{R}.\widehat{R}^* = 1$ and $\widehat{L}.\widehat{R}^* = 0$. The two polarizations are therefore orthogonal. The field $\vec{E}$ can thus be written:

$$\vec{E} = E_L\left(\widehat{L} + \frac{E_R}{E_L}.\widehat{R}\right) = E_L\left(\widehat{L} + \rho_c.\widehat{R}\right) \qquad [2.75]$$

where:

$$\rho_c = \frac{E_R}{E_L} = |\rho_c|.e^{j\delta_c}$$

Elliptical polarization is the combination of two circular polarizations: one right-hand, with radius A, and the other left-hand, with radius B, or vice versa, as shown in Figure 2.8.

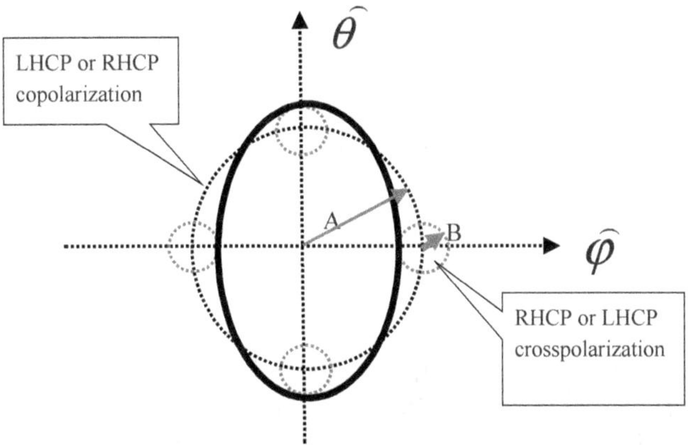

Figure 2.8. *Elliptical polarization*

The larger axis of the ellipse is formed when the two polarizations are in phase. The smaller axis occurs when the two polarizations are in opposite phase. The axial ratio (RE) is given as the ratio between the two axes:

$$RE = \frac{A-B}{A+B} \qquad [2.76]$$

$A = E_L$, $B = E_R$ or vice versa.

In the case of frequency reuse, it is specified that $RE = 1.06$, so $RE\left(dB = 20.\log_{10}\left(1,06\right)\right) \approx 0.5\text{ dB}$, which corresponds to decoupling $\frac{A}{B} = \frac{1-RE}{1+RE}$ between the two polarizations, which, in decibels, is ≈ 30.7 dB.

2.9. Calculating the electromagnetic field radiated by an antenna[15]

2.9.1. *Expanded discussion of the EFIE and MFIE formulae*

Looking again at the rigorous equations [2.47a] and [2.47b], with all the intermediary calculations having been performed, we obtain:

$$\vec{E}(\vec{r}) = \frac{\eta.k^2}{4.\mu}\iiint_V \begin{bmatrix} \vec{J}(\vec{r}).\left(-\frac{j}{k.R} - \frac{1}{k^2.R^2} + \frac{j}{k^3.R^3}\right) + \\ \left[\vec{J}\left(\vec{r}'\right).\hat{u}\right].\hat{u}.\left(\frac{j}{k.R} + \frac{3}{k^2.R^2} - \frac{3.j}{k^3.R^3}\right) \end{bmatrix} .e^{-j.k.R} dv \quad [2.77a]$$

$$\vec{H}(\vec{r}) = \frac{k^2}{4.\pi}\iiint_V \vec{J}\left(\vec{r}'\right) \wedge \hat{u}.\left(\frac{j}{k.R} + \frac{1}{k^2.R^2}\right).e^{-j.k.R}.dv \quad [2.77b]$$

This formulation applies to all types of antennas, whether larger or smaller than the wavelength. It is used to calculate the EM field as a function of the distance from the antenna.

The currents appearing in equations [2.77a] and [2.77b] are the currents over the surfaces of structures in the volume V such as those shown in Figures 2.3, 2.9a and 2.9b. At large distances, these antennas, whatever their size, are treated as single points in the volume *V*. In addition, to facilitate certain end-to-end calculations numerically, we employ the equivalence theorem to replace the currents circulating across these structures with the fictitious tangential fields or electrical and magnetic currents across a surface which contains the antenna (Figures 1.14a and 1.14b), for example. The surface may be open and flat (Figure 1.14a), as is the case with the aperture method (Silver 1984).

15 Milligan 2005; Balanis 2008; Huang 2008.

a)

b)

Figure 2.9. *a) Spiral antenna in the L band. Printed lines $\lambda / 2$ circular polarization; b) ground station antenna with diameter D = 510 λ in the C band, with frequency reuse in LHCP and RHCP (source: Pleumeur-Bodou photo PB4)*

2.9.2. *Calculations for an elementary dipole*

It is only possible to analytically solve equations [2.78a] and [2.78b] with simple so-called *reference cases*. Without sacrificing generality, with the superposition theorem, these elementary antennas can enable us to numerically calculate complex antennas by sampling.

The reference elementary antenna, here, is an elementary source with length Δl and constant intensity I along the $\hat{z}$ axis. When $\Delta l << \lambda$, we see quasi-static operation far from the occurrence of the phenomena of resonance, which holds true all the more when the frequency is low. Thus, from the standpoint of the configuration of the fields, the radiating system behaves like a low-pass filter, having a response that is *frequency-independent* from a frequency of zero up to a range near to the occurrence of the first resonance for $\Delta l \approx \frac{\lambda}{4}$:

$$\vec{J} = \hat{z}.I.\Delta l \qquad [2.78]$$

This current element is electrically short $(\Delta l << \lambda)$ and is fictitious. By replacing $\vec{J}$ in equation [2.81a], we find (Huang 2008), in the Cartesian coordinates system shown in Figure 2.1:

$$E_r = 2.\frac{I.\Delta l}{4.\pi}.\eta.k^2.\cos(\theta).\left(\frac{1}{k^2.R^2} - \frac{j}{k^3.R^3}\right).e^{-j.k.R} \qquad [2.79a]$$

$$E_\theta = \frac{I.\Delta l}{4.\pi}.\eta.k^2.\sin(\theta).\left(\frac{j}{k.R} + \frac{1}{k^2.R^2} - \frac{j}{k^3.R^3}\right).e^{-j.k.R} \qquad [2.79b]$$

$$E_\varphi = 0 \qquad [2.79c]$$

η is the medium's intrinsic impedance.

Thus, we have:

$$H_r = 0 \qquad [2.79d]$$

$$H_\theta = 0 \qquad [2.79e]$$

$$H_\varphi = \frac{I.\Delta l}{4.\pi}.k^2.\sin(\theta).\left(\frac{j}{k.R} + \frac{1}{k^2.R^2}\right).e^{-j.k.R} \qquad [2.79f]$$

Note that for two components E_θ and E_φ, we have one component H_φ.

We can see that in the near zone, the electric field has a radial component E_r, which decreases in $\frac{1}{R^2}$, while the other component E_θ varies in $\frac{1}{R}$ with the magnetic field H_φ. The null radial component is the so-called *radiation condition.*

In the far zone (the Fraunhofer region), when $k.r >> 1$ – i.e. when $r >> \frac{\lambda}{2.\pi}$, $\frac{1}{k.R} >> \frac{1}{(k.R)^2} >> \frac{1}{(k.R)^3}$.

We obtain:

$$E_r \approx 0 \quad [2.80a]$$

$$E_\theta = j.\frac{I.\Delta l}{4.\pi}.\eta.k.\sin(\theta).\frac{e^{-j.k.R}}{R} \quad [2.80b]$$

$$E_\varphi = 0 \quad [2.80c]$$

For the magnetic field:

$$H_r = 0 \quad [2.80d]$$

$$H_\theta = 0 \quad [2.80e]$$

$$H_\varphi = j.\frac{I.\Delta l}{4.\pi}.k.\sin(\theta).\frac{e^{-j.k.R}}{R} \quad [2.80f]$$

Now we have:

– a component for the electric field and a component for the magnetic field;

– both fields are inversely proportional to the distance R;

– both fields are orthogonal, as stated above;

– the Poynting vector:

$$\vec{S} = \vec{E} \wedge \vec{H}^* = \widehat{R}\left(\frac{I.\Delta l}{4.\pi.R}.k.\sin(\theta)\right)^2.\eta \quad [2.81]$$

The Poynting vector is maximum for $\theta = \frac{\pi}{2}$ in the plane $< \widehat{x}, \widehat{y} >$.

The condition $R >> \frac{\lambda}{2.\pi}$ is introduced for electrically short antennas (see section 3.8). It is simply a function of the frequency, not connected to a dimension D

of the antenna. When D is electrically long ($D > \lambda$), the condition of $R > \dfrac{2.D^2}{\lambda}$ is the common definition of the far field, which corresponds to a phase error of $\pi / 8$ in receiver mode at any given point on an antenna.

If the antenna is not considered electrically long, the recommended condition for the far region is $R > 3.\lambda >> \dfrac{\lambda}{2.\pi}$. Certain authors use $R > 10.\lambda$, or other conditions that ensure the wave is plane beyond the distance in question. Thus, the different regions are not precisely fixed; the boundaries between them are fuzzy. For example, for a dipole $D = \lambda / 2$, we have $\dfrac{2.D^2}{\lambda} = 0.5.\lambda$, which is too small a distance to be considered a far field.

The distance $R < \dfrac{\lambda}{2.\pi}$ is called the *reactive zone* of the EM field, which rapidly changes with distance. This reactive zone contains a real part of the energy being propagated, representing the resistance to radiation R_A (equation [1.2]), and contains an imaginary part X_A of the energy which remains stored around the antenna, representing the reactance of the impedance $Z_A = R_A + jX_A$ of the antenna.

Here, we shall not discuss *loop* antennas, which are the second category of the most common antennas. The expressions are close to those for the elementary electric dipole described earlier; they are also known as *elementary magnetic dipoles*. In the vicinity of the antenna (the near-field zone), the magnetic field is predominant, which is why it is used for applications in radiofrequency identification (RFID).

2.10. Aperture antenna[16]

2.10.1. *Wireless radiation of apertures*

When the dimensions of the antennas are fractions of the wavelength, or alternatively are much larger than the wavelength, as shown in Figures 2.9a and 2.9b, and particularly when the constitutive elements are electromagnetically coupled, an analytical solution cannot be found for a rigorous result, throughout the space, at any given distance. One of the solutions to solve this problem is to

16 Drabowitch 1978; Silver 1984; Milligan 2005; Balanis 2008.

calculate the EM field on a closed surface (in the best case) by applying the equivalence theorem. For antennas such as horns or reflector antennas, this surface Σ is placed opposite their aperture. It is often considered plane to simplify the calculations, and limited to the diameter of the aperture (Figure 1.15a) for a horn, a waveguide or a Cassegrain antenna, etc.

We suppose the aperture to be sufficiently wide for the EM wave radiated by the structure to be a plane wave on Σ and such that the effect of the current lines on the edge of the structure is negligible on Σ. The field radiated by such an aperture, in keeping with the Kirchhoff–Fresnel approximation, manifests as the Fourier transform of the illumination law with a quadratic phase factor. We know that an eigenfunction of the Fourier transform is a Gaussian function. Consequently, the radiated field at any point by Gaussian illumination itself assumes Gaussian form. In addition, a Gaussian beam is a specific solution to the Helmholtz propagation equation, as is a plane wave under the paraxial approximation. This result is at the heart of the possibility of energy transport by a repeated focusing system (Figure 2.9b). Indeed, if it is possible to find, before the aperture, a plane where the distribution of the field in amplitude has the *same variance* as on the aperture, then all we need is to set a focalizer system in that plane, also re-establishing the quadratic phase law to return to the initial conditions. By recurrence, we can thus establish a chain of focalizers to transport the energy over a distance which, in practice, is unlimited, with negligible losses. This principle is exploited for large antennas for ground stations with a *beam waveguide feed*, as shown in Figure 2.9b. For an aperture, aside from the main lobe, all other lobes are non-Gaussian. The maximum angle of divergence generally accepted, in which the pattern remains in a Gaussian variation, is between 20° and 30°, which gives a cone of between 40° and 60°. The complex electric field of a Gaussian beam measured (in volts/m) at the distance r from the center of the beam at the distance z from its origin is:

$$E(r,z) = E_0 \frac{w_0}{w(z)} e^{-\frac{r^2}{w^2(z)}} e^{-j\left(k.z + k\frac{r^2}{2R(z)}\right)} \qquad [2.82]$$

In a plane with a straight section, the average temporal intensity (or radiance), measured in watt/m^2, is:

$$P(r,z) = \frac{|E(r,z)|^2}{2Z_0} = E_0^2 \left(\frac{w_0}{w(z)}\right)^2 e^{-\left(\frac{2.r^2}{w^2(z)}\right)} \qquad [2.83]$$

with:

– $k = {}^{2\pi}/_{\lambda}$ the wavenumber in free space;

– $w(z)$ the distance to the center of the beam on the axis, at the point where the amplitude of the electric field E is multiplied by $1/e = 0.3679$, which corresponds to an energy decrease by $\left(\frac{1}{e}\right)^2 = 0.1353\ \text{watts/m}^2$ (Figure 2.10). This parameter is called the *beamwidth*;

– E_0 the amplitude of the beam on the axis $E_0 = E(0,0)$;

– Z_0 the impedance in a vacuum $Z_0 = 120\pi = 377\ \Omega$.

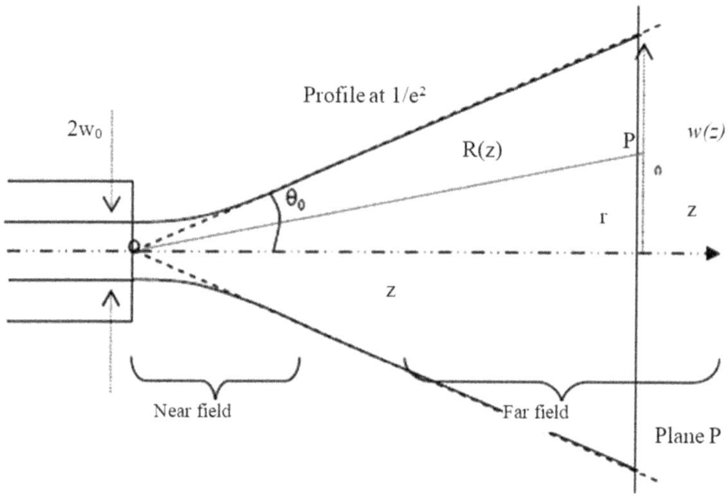

Figure 2.10. *Evolution of the EM field on exiting a radiating aperture*

At the output from the radiating aperture of a guide, a horn or a parabola, the beam will display the following behavior: the half-width $w(z)$ will have its minimal value w_0 at the origin (Figure 2.10). For a wavelength λ and at a distance z along the axis of the beam, the variation in beamwidth will be:

$$w(z) = w_0 . \sqrt{1 + (\frac{z}{z_0})^2} \quad [2.84]$$

where the origin of the z axis is defined, without sacrificing generality, as the point of origin:

$$z_0 = \frac{\pi . w_0^2}{\lambda}$$

z_0 is known as the Rayleigh range or depth of field.

The radius of curvature of the wavefront sphere is closely approximated by:

$$R(z) = z.[1+(\frac{z_0}{z})^2]$$

At a distance from the origin equal to z_0, the beamwidth is $w(\pm z_0) = w_0.\sqrt{2}$.

The distance at $+z_0$ and $-z_0$ is known as the confocal parameter:

$$b = 2.z_0 = \frac{2.\pi . w_0^2}{\lambda}$$

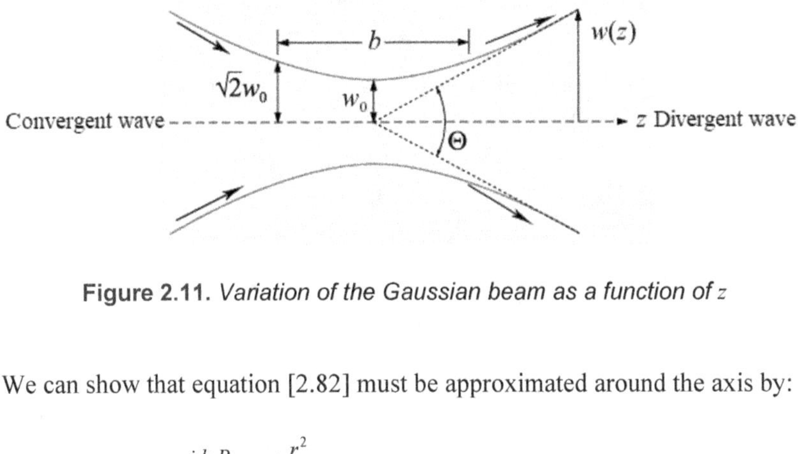

Figure 2.11. *Variation of the Gaussian beam as a function of z*

We can show that equation [2.82] must be approximated around the axis by:

$$E(r) = E_0 \frac{e^{-j.k.R}}{R} e^{-jk\frac{r^2}{2R}} \qquad [2.85]$$

Equation [2.85] is compliant with the radiation at any given distance from an aperture obtained with the Kirchhoff formulas in the Fresnel approximation (Drabowitch 1978), where the quadratic phase term is the expression of the transfer from point M to point Q (MQ) as illustrated in Figure 2.12.

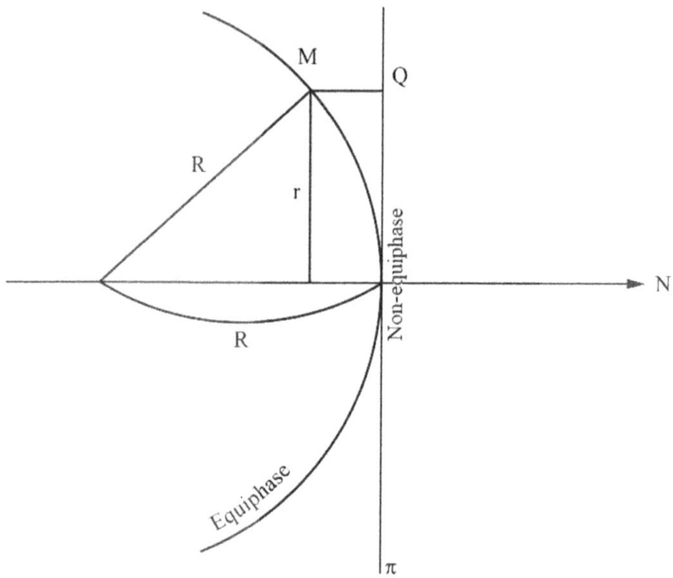

Figure 2.12. *MQ is the equiphase correction on the sphere to the quadratic term on the plane*

To infinity, on the sphere whose radius of curvature is R, close to the axis, the amplitude of the field is approximated by:

$$E(r) = E_0 \frac{e^{-j.k.R}}{R} \quad [2.86]$$

In the case of a centered system in receiver mode, Figure 2.11 represents the convergence of the Poynting vector toward the focal plane situated where the beam is at its narrowest, followed by its divergence. This curve enables us to position a horn in the focal plane and define it as *optimum* in terms of energy transfer. Note that in geometrical optics (GO), the convergence is reduced to a single point (the focal point), unlike the Airy disc in physical optics (PO).

2.10.2. *Identification of the different zones*[17]

It is fundamentally important to know the power level radiated by an antenna as a function of its distance for at least three reasons:

17 Silver 1984; Huang 2008.

1) Within an array of ground stations, this knowledge will allow us to manage the positioning of the antennas on site by calculating the electromagnetic coupling that occurs between them; to calculate the radiation patterns in compliance with the ITU-R recommendations.

2) The near-field calculation enables us to evaluate the SAR (specific absorption rate) in W/kg, which is an indicator of the amount of energy carried by the radiofrequency waves received by the user of a mobile phone/handset. The installation of base stations in an urban environment is highly sensitive to the population – hence EU Directive 2013/EU on protection of workers. In addition, AFSSET (*Agence française de sécurité sanitaire de l'environnement et du travail* – French Environmental and Occupational Health and Safety Agency), as a precautionary measure, limits handset power for a monitored absorption rate for human tissue:

$$SAR = \frac{\sigma.\left|\vec{E}\right|^2}{2.\rho} = \frac{\left|\vec{J}\right|^2}{2.\sigma} \text{ in (W/kg)}$$

where σ is the conductivity of the tissue (S/m), ρ the density of the tissue (kg/m^3) and $\left|\vec{J}\right|$ the current density in the tissue (A/m^2).

The European standards (Directive 1999/5/EC) define recommendations for the 10 Mhz–300 GHz range, based on a full-body SAR < 0.08 W/kg.

3) It allows us to qualify measurement bases as *near-field* or *far-field.*

2.10.2.1. *Near and far fields*

For a circular aperture, Figure 2.13 shows the evolution of the radiation pattern of a circular plane aperture for an equiamplitude and equiphase distribution of the electric field across Σ.

Note that, only if the phase of illumination of the aperture does not vary rapidly in comparison to λ, then the strict radiation pattern generated by an integral equation, in the far field, has an analytical solution given by the Fourier transform of the illumination on the aperture.

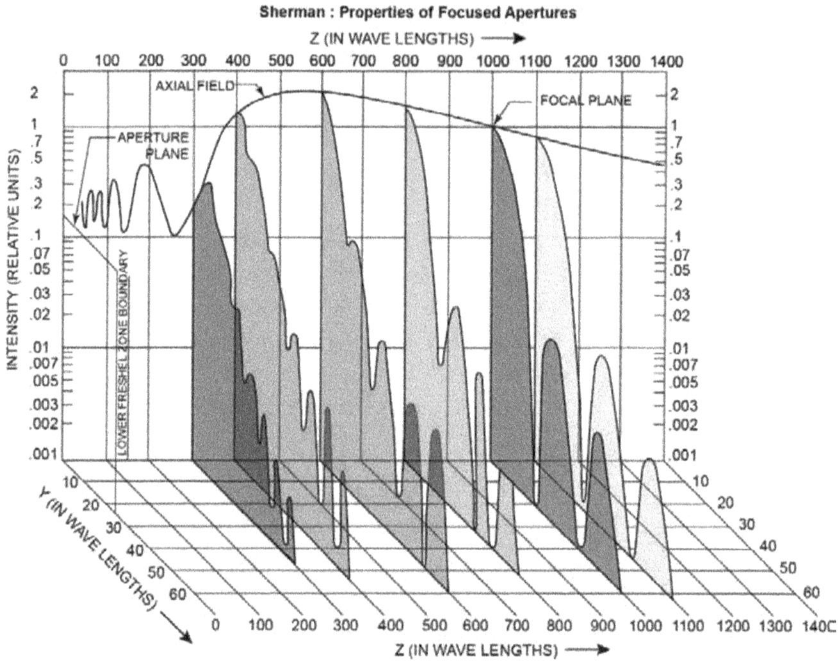

Figure 2.13. *Radiation pattern as a function of distance from a circular aperture with constant amplitude and phase*

2.10.2.2. *Near radiation zone (Rayleigh zone)*

For a circular aperture with diameter *D*, at a distance *R* such that:

$$R < 0.62\sqrt{\frac{D^2}{\lambda}}$$

the EM field is reactive. In fact, it is the transition zone imposed by the boundary conditions on conductive surfaces for the switch from evanescent modes to EM field propagation modes. As indicated above, it represents the imaginary part of the impedance of the antenna. Except when we are dealing with electrically short antennas, this reactive part is negligible.

The Rayleigh zone extends from the aperture to a distance $R = \frac{D^2}{2\lambda}$. It is defined as being the distance where the phase term $k\frac{r^2}{2R}$ in equation [2.85] represents a phase error MQ of $\frac{\pi}{2}$ (Figure 2.12):

$$\Delta\phi = \frac{2\pi}{\lambda}\frac{D^2}{8R} = \frac{\pi}{2}, \text{ so } R = \frac{D^2}{2\lambda}$$

In this cylinder, the wave is still plane.

2.10.2.3. *Intermediary zone (Fresnel zone)*

The Fresnel zone, which sometimes encapsulates the Rayleigh zone, in principle, extends from the zone $R = \frac{D^2}{2\lambda}$ to the zone $R = \frac{2D^2}{\lambda}$ – that is, four times the range of the Rayleigh zone. The distance $R = \frac{2D^2}{\lambda}$ corresponds to a maximum phase error of $\frac{\pi}{8}$ on the edge of the aperture (Figure 2.14):

$$\Delta\phi = \frac{2\pi}{\lambda}\frac{D^2}{8R} = \frac{\pi}{8}, \text{ so } R = \frac{2D^2}{\lambda}$$

In this zone, the wave already has the structure of a spherical wave, whose amplitude varies significantly, as shown in Figure 2.13.

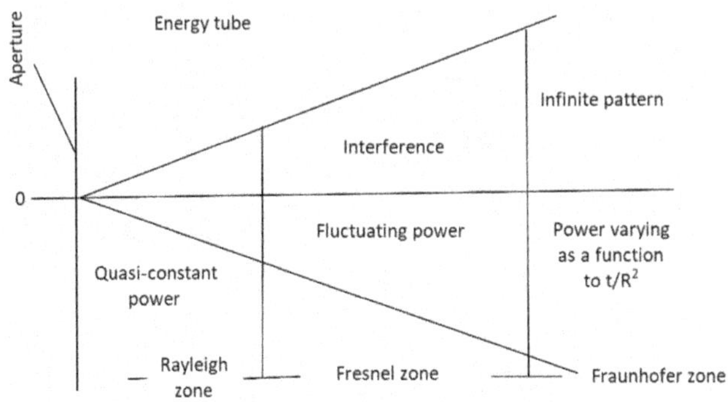

Figure 2.14. *The different zones*

2.10.2.4. *Far field (Fraunhofer zone)*

In the Fraunhofer zone, quantitatively from $R = \dfrac{2D^2}{\lambda}$, the wave is spherical and said to be *locally plane*, in view of the variation of its amplitude in $1/R$. The energy varies in $1/R^2$ in a monotonic fashion (Figure 2.13). From that distance on, the pattern is formed. We can therefore remove this latter, without correction with a measurement base whose transmitter or receiver is situated at around this distance and beyond.

If, for example, $D = 10$ meters, $\lambda = 0.01$ m at 30 GHz, we have:

$$R = \frac{D^2}{2\lambda} = 5{,}000 \text{ meters}$$

$$R = \frac{2D^2}{\lambda} = 20{,}000 \text{ meters}$$

which demonstrates the limit of the possibilities of measure with so-called *long* bases.

3

Different Antenna Technologies

Jean-Pierre BLOT
EUROSATCOM, Les Ulis, France

3.1. Horns[1]

Horns are extremely important in numerous spatial and terrestrial applications. Their frequency bandwidth is far greater than that of any other technology, such as wire antennas or printed antennas. They also have greater directivity. Above all, they radiate a pattern with very pure polarization, which cannot be achieved with other technologies, for a reasonable cost. In addition, when they are made from highly conductive metals (e.g. copper, gold, silver, etc.), the ohmic losses incurred may be negligible (of the order of 0.001 dB), with a reflection coefficient (vswr) which, in practice, is less than 1.05 on average, meaning they can transmit tens of kilowatts of power without any time constraints. In the receiver band, the noise temperature produced by the ohmic losses is slight, or even negligible, which is extremely important in space applications and deep-space probes, for example.

A horn is made up of a guide bar with a widening cross-section, finishing in a radiating aperture. The increasing progression from the meridian has a twofold advantage: it enables large bandwidth and the gradual passage of the impedance from free space to that of the power supply guide, and vice versa (Figure 3.1). This type of antenna is very widely used as a primary source, and offers advantages on

For a color version of all figures in this book, see www.iste.co.uk/favennec/electromagnetic2.zip.

1 (Drobowitch 1978; Silver 1984; Balanis 2008; Huang 2008).

both the mechanical and radioelectric fronts. It is easy to protect it from bad weather with a radome closing the aperture, and transparent to radiation. Surplus pressure inside prevents the condensation of water vapor, and improves the power handling. In addition, very diverse radiation characteristics may be obtained depending on the type of horn chosen, and on its excitation conditions: polarization and propagation modes.

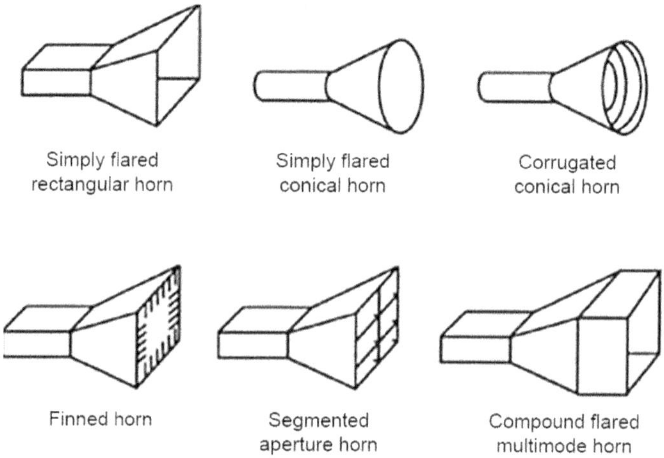

Figure 3.1. *A few examples of horns*

Indeed, there are several different categories of horns: single-mode, multi-mode and corrugated. In addition, in each category, the horns differ in terms of their geometry: angle of aperture and shape of a transverse section. To assess the radiation characteristics, we can often make the hypothesis that a horn is equivalent to a radiating aperture whose illumination stems from the mode, or modes, of propagation excited in the guide to its mouthpiece. Sometimes, we need to take account of the existence of higher modes of varying origins: discontinuity formed by the aperture itself, parasitic reflections, etc. However, if the horn is fed by a guide transporting its fundamental mode, the longitudinal profile of the horn provides a gradual transition to the aperture and prevents the appearance of parasitic higher modes. The illumination obtained is then the faithful image of the distribution of the field of the mode in question (single-mode horn). On the other hand, if the horn exhibits discontinuities, higher modes are created which may be desirable – for example, in the case of the Potter horn. For this horn, the calibrated addition of the TM11 mode to the fundamental mode TE11 renders the radiation pattern symmetrical by a mode close to the hybrid mode HE11 (Figure 3.2), like for corrugated horns, but at a lower cost and much lower frequency band.

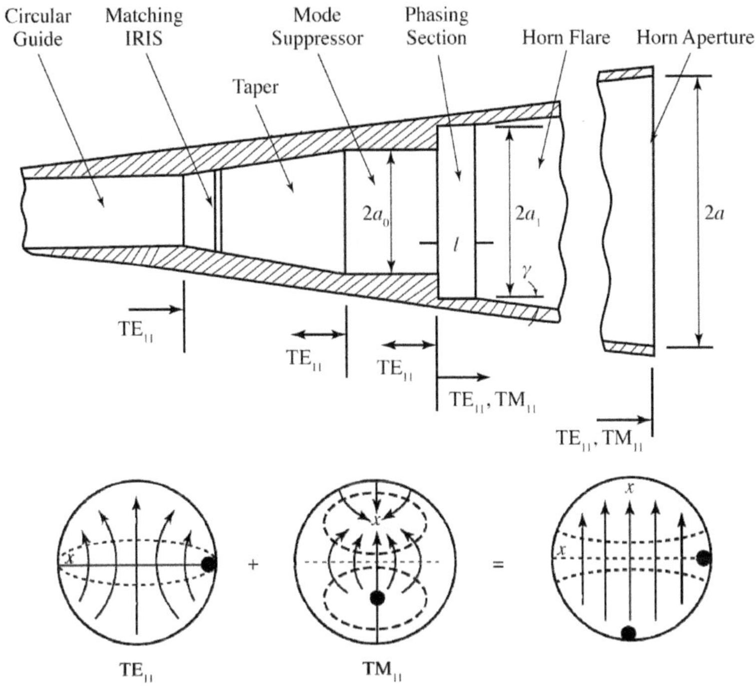

Figure 3.2. *Potter bimode horn*

A horn may be fed by multiple input guides (or *accesses*). This results in multiple radiation patterns. If the arrangement of the horn is such that the various accesses are not coupled together, the radiated patterns are orthogonal. In relation to receiving, in the case of single-mode horns, the diffraction pattern, in the vicinity of the aperture, engenders a set of diverse order modes (Rayleigh zone), which are liable to propagate toward the mouth of the input guide, up until the planes whose cross-section corresponds to the cutoff of the modes in question. Finally, only the useable mode remains. Thus, the horn acts as a mode filter. For further details, refer to the published works cited in the References.

3.2. Coaxial cables and input guides in antennas[2]

Antennas are fed by coaxial cables and/or by metal waveguides. A metal guide (Figure 3.3) is fed by coaxial probes, or slots, because of the compatibility of the

2 (Huang 2008; Geyi 2015).

EM field of the probe, or slots, with the guide's fundamental mode. In any event, there is always a coaxial terminal at a certain point.

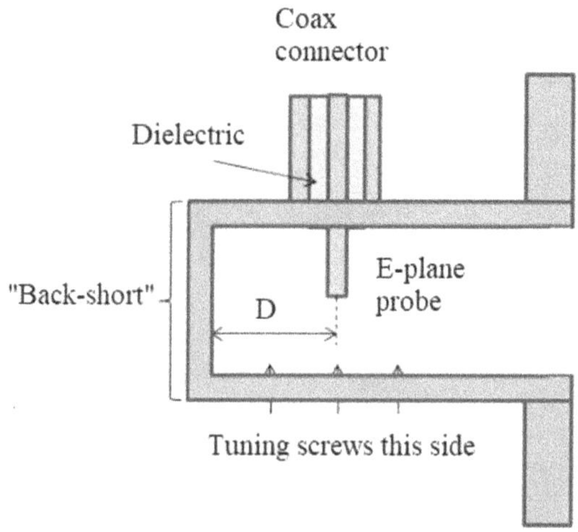

Figure 3.3. *Example of a coaxial cable to rectangular/circular waveguide transition*

In general, multiple types of electromagnetic guides exist, depending on the frequencies that we wish to transport, the performances in terms of bandwidth, the power that we wish to transmit, and low reception losses. Figure 3.4 illustrates the main base technologies.

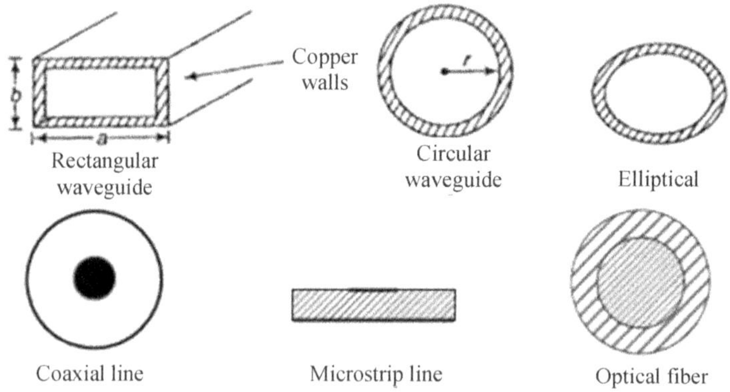

Figure 3.4. *Base waveguides*

Waveguides may be rectangular or square, smooth circles or grooved, partially or completely filled by a dielectric with a variable constant ε but small ohmic losses or metal ridges of varying thicknesses to expand the bandwidth, depending on the use being made of the equipment, as illustrated in Figure 3.9 (section 3.2.2.1).

Hollow metal waveguides are at the root of the input for reflector antennas, which are required to deliver excellent performance in terms of transmitted power and noise temperature on receiving.

Flexible or semi-rigid cables are systematically used in equipment cabinets as connecting lines between two hyperfrequency devices. They are used for input to wire antennas.

Flat technologies such as microstrips are found in multiple variants, depending on low-power use, such as for MMIC (Monolithic Millimeter-Wave Integrated Circuits) integrated technologies. Similarly, to supply printed technology antennas (patch antennas, etc.). These supply lines are quickly dispersive, and are often protected by a shield (such as shielded microstrip line technology).

The optical fibers are not connected to the antenna directly, but rather through a transducer situated upstream of the antenna's power supply circuit. They are used when very long distances need to be covered: ranging between a few kilometers and hundreds of kilometers before the signal is regenerated and amplified by *repeaters*. They are gradually coming to replace coaxial lines in technical cabinets.

3.2.1. *Coaxial cables*

Compared to a standard cable, for a reasonable additional cost, a coaxial cable of greater quality, in terms of bandwidth and ohmic losses, can be used for connections over tens of meters between a central building and an antenna. Such a cable, in the case of low transmission power, can be substituted for a hollow ridged metal elliptical guide, which is highly costly and tricky to install (Figure 3.6c). Standard coaxial cables, which generally have a characteristic impedance of 75 Ω, have *general consumer* applications, such as the link between a Yagi-Uda-type antenna (indoor or outdoor) and a television set.

A coaxial cable propagates the TEM mode (Figure 3.5) up to a high cutoff frequency f_c of the TE11 mode of several GHz, depending on the diameter $d_1 = a$ of the central conductor and the internal diameter $d_2 = b$ of the shield and the dielectric. The LHI dielectric is non-magnetic and organic, such as polyethylene. The relative permittivity is approximately $\varepsilon_r \approx 2$. The prevailing mode of propagation is the TEM mode, which occurs from 0 Hz up to the cutoff frequency f_c of the TE11 mode, given by (Pozar 1997):

$$0 < f < f_c = \frac{c}{\pi.(a+b).\sqrt{\varepsilon_r}} \qquad [3.1]$$

where $c \approx 3.10^8$ m/s, the celerity of light.

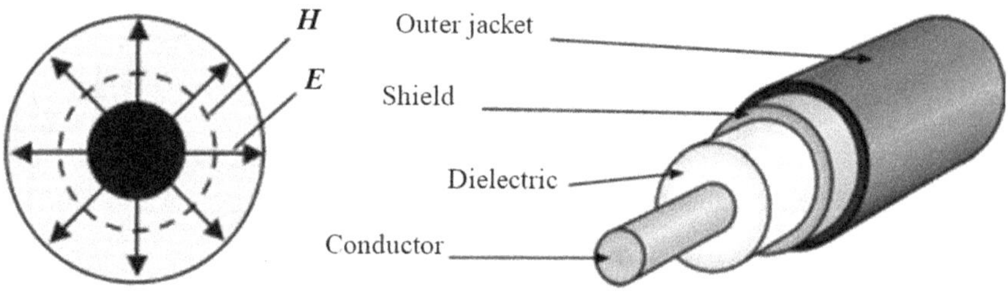

Figure 3.5. *Electric field and magnetic field lines of f_c mode in TEM mode*

As shown by equation [3.1], to drive back the propagation of TE11 mode as far as possible, the circumference of the shield on the coaxial cable must be smaller than the wavelength λ, which limits the use of coaxial cables to very high frequencies – for technical reasons of feasibility of very thin cables – making them costly and fragile.

Above the cutoff frequency, the cable propagates undesirable higher modes, giving rise to rapidly increasing losses. These losses, for a so-called *low-loss cable*, are given by:

$$\alpha \approx \sqrt{f}.\frac{\sqrt{\mu.\varepsilon_r}.\left(\frac{1}{a}+\frac{1}{b}\right)}{120.\sqrt{\pi.\sigma_2}.\log_e\left(\frac{b}{a}\right)} + \frac{60.\pi.\sigma_1}{\sqrt{\varepsilon_r}} \qquad [3.2]$$

The characteristic impedance Z_0 is given where $\mu_r = 1$ by:

$$Z_0 = \frac{60}{\sqrt{\varepsilon_r}}.\log_e\left(\frac{b}{a}\right) \ (\Omega)$$

With $\frac{b}{a} \approx 3.592$, the impedance is 76.7 Ω, and the attenuation reaches its minimum. The standard impedance of these cables is typically 50 Ω or 75 Ω.

The power of a circuit signal, whether instantaneous or average, depends directly on the voltage and current flowing through the conductors. For the coaxial cables represented in Figure 3.5, the power that it is possible to inject is of the order of a few tens of watts, as a function of the *b/a* ratio. This depends on the quality of the metal conductors (usually copper), and on the quality of the dielectric, which is usually a polymer such as polytetrafluoroethylene (PTFE). With pressurized coaxial cables, the breakdown voltage between the two conductors is approximately 30 kV/cm. Common coaxial cables are used in the domain of antennas up to just over 20 GHz, following the previous technical observation, and with losses that are incompatible for highly professional applications. They are placed before a high-frequency transposition in a *transmission chain*, before the power amplifier to the antenna. In the *receiver chain*, they are found downstream of the antenna, generally after a low-frequency transposition and a low-noise amplifier (LNA) at the top of the downlink. In the case of *microstrips*, as illustrated in Figure 3.4, the mode which is propagated is a quasi-TEM mode because of the open ground plane $E_z \neq 0, H_z \neq 0$, but $|E_z| << |E_{tranverse}|$ and $|H_z| << |H_{tranverse}|$. The possible powers are those of the integrated MMICs – i.e. a few milliwatts.

3.2.2. *Waveguides*

Metal waveguides (Figures 3.6b, 3.6c and 3.6d), assimilated to electromagnetic transmission lines, are unique. No other transmission lines are made of a single metal piece: usually tubular, rectangular, square, circular or elliptical. A hollow metal waveguide such as those shown in Figure 3.6, due to the boundary conditions for the electric field and the magnetic field on metal walls and the satisfaction of Maxwell's equations, means that a multitude of modes can propagate there. They are *transverse electrical (m, n)* modes $TE_{m,\ n}$ and *transverse magnetic (m, n)* modes $TM_{m,\ n}$.

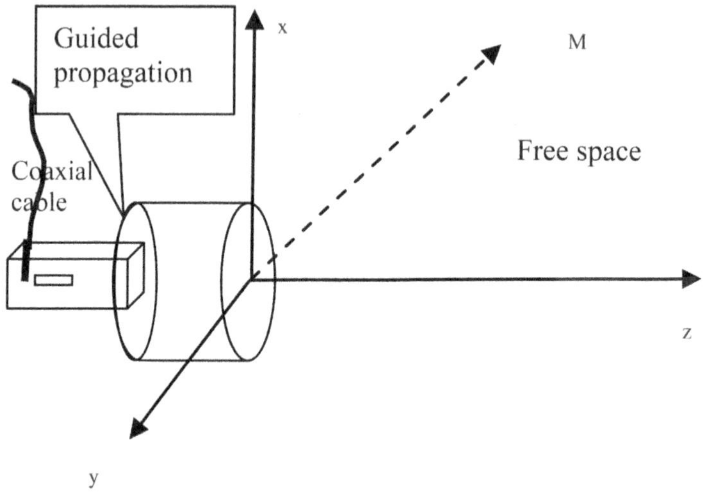

Figure 3.6a. *Guided and free propagation*

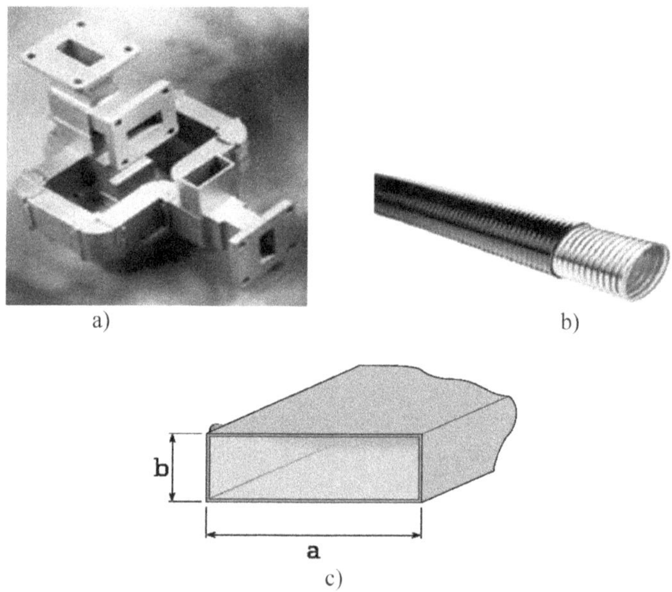

Figure 3.6b. *a) Example of orthomode transduction or OMT; b) corrugated elliptical guide; c) rectangular guide*

3.2.2.1. *Rectangular or square guides*

For the rectangular guide, the modes are generated by a longitudinal component (following the direction of propagation). For TE modes: $Ez = 0$ and $Hz \neq 0$. For TM modes: $Ez \neq 0$ and $Hz = 0$. The TEM mode has no longitudinal component; its cutoff frequency is 0 Hz, so it cannot exist in such guides.

In the case that a rectangular guide feeds a horn, in view of the electric field lines of the fundamental mode $TE_{1,0}$, the rectangular guide means that we can use a single form of polarization along the axis of the short side of the guide (Figure 3.7), which restricts the applications but is interesting, without the horn, as a measurement probe. For applications in telecommunications, using only one type of polarization means sacrificing the possibility of reusing the frequency band, which enables a second polarization to be used. Noting that the fundamental mode $TE_{1,0}$ is independent of the short side of the guide, we can either reduce that side to limit the bulk and create a compact device, such as an orthomode (Figure 3.6b) in the proportions $a \approx \lambda/2$ and $b \approx \lambda/4$, or obtain a square guide $a = b$, set between the orthomode and the horn to obtain the two polarizations. The orthomode supplies the square guide with two or four reduced guides, each terminating in a slot adapted on two faces or on the four faces of the square guide to optimize the decoupling of polarization by geometric symmetry. In the same way, the square guide can be fed by two or four coaxial cables, as shown in Figure 3.3.

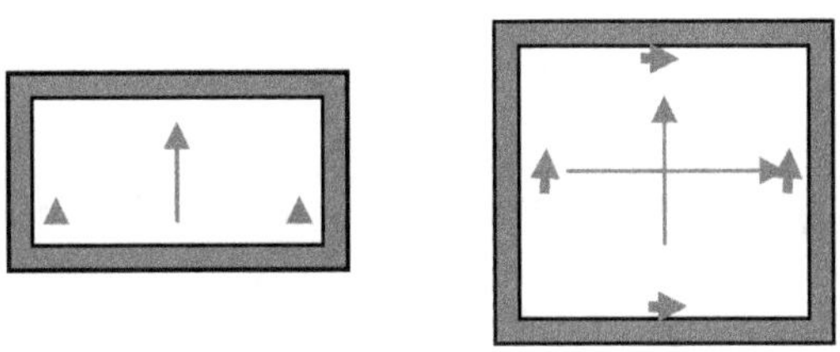

Figure 3.7. *With a square guide, we can reuse frequencies by orthogonal polarization*

The TE and TM modes are treated in the same way, with the exception that for TM modes, the values (0, *n*) and (*m*, 0) must be excluded as they yield a null longitudinal field ($E_z = 0$), whereas with TE modes, anything is possible apart from (0,0).

When $a \geq b$, the minimum cutoff frequency is that of the $TE_{1,0}$ mode. This mode is referred to as fundamental because it is the only one to propagate up to the cutoff frequency of the following mode, the TM_{01} or TE_{20} modes, which makes the propagation unique: the propagation is therefore not dispersive in the intermodal sense (all the energy is concentrated in a single mode).

When we know the dimensions *a* and *b*, we can classify the modes by increasing cutoff frequency $f_{c(m,n)}$, and thus know when a mode, whether desired or otherwise, may appear in a determined frequency band. For example, in the case of reflector antennas, the fundamental mode TE_{10} will be used to transport telecommunications information, whereas odd-numbered modes TM_{01} and/or TE_{21} will be used to generate the signal to orientate the antenna toward the satellite as tracking modes.

The benefit of using the fundamental mode $TE_{1,0}$ for transporting usable energy is that, as previously indicated, its field lines are compatible with those created by dipole-type elementary probes or by slots positioned along the current lines.

The cutoff frequency of the TE and TM modes:

$$f_c = \frac{v}{2}\sqrt{(\frac{m}{a})^2 + (\frac{n}{b})^2} \qquad [3.3]$$

for b < a < 2b $v = \frac{1}{\sqrt{\mu.\varepsilon}}$ the propagation rate.

The attenuation is:

– for the $TE_{m,n}$ modes:

$$\alpha_{TEmn} = \frac{2.R_S}{b.\eta.\sqrt{1-\left(\frac{f_{cmn}}{f}\right)^2}}\left\{\left[1+\frac{b}{a}\right]\left(\frac{f_{cmn}}{f}\right)^2 + \frac{b^2m^2+a.b.n^2}{b^2m^2+a^2n^2}\left[1-\left(\frac{f_{cmn}}{f}\right)^2\right]\right\} \text{(Np/m)}$$

– for the $TM_{m,n}$ modes:

$$\alpha_{TMmn} = \frac{2.R_S}{b.\eta.\sqrt{1-\left(\frac{f_{cmn}}{f}\right)^2}}\left\{\frac{b^3m^2+a^3.n^2}{a.b^2m^2+a^3n^2}\right\} \text{(Np/m)}$$

with the surface resistance $R_s = \sqrt{\dfrac{\pi . f . \mu_0}{\sigma}}$.

As with a coaxial cable, the higher the frequency, the greater the losses, in view of the coefficient $\sqrt{f}$. To limit the noise temperature in the receiver band for satellite applications, the inside of the guides is often coated with silver or gold, which limits ohmic losses as far as possible.

The guided wavelength far from the cutoff frequency is given by:

$$\lambda_g = \frac{\lambda_0}{\sqrt{1 - \dfrac{\lambda^2_0}{\lambda^2_c}}} = \frac{\lambda_0}{\sqrt{1 - \dfrac{\lambda^2_0}{4}\left(\dfrac{m^2}{a^2} + \dfrac{n^2}{b^2}\right)}} \qquad [3.4]$$

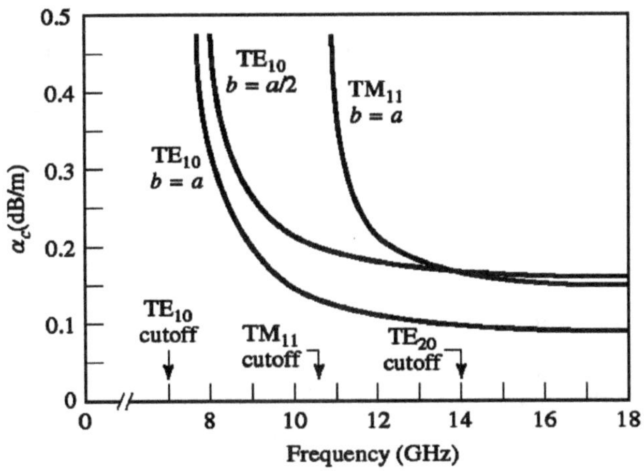

Figure 3.8. *Example of losses as a function of the a/b ratio above 7 GHz*

The guided wavelength $\lambda_g > \lambda_0$ is the wavelength in a vacuum.

The bandwidth of the guide is considered to be that extending from the cutoff frequency of the mode in question, to the cutoff frequency of the next mode. Beyond that point, there is a risk of the appearance of undesirable modes and thus modal dispersion.

Figure 3.9. *Ridged guides*

Ridged guides, for example, allow us to reduce the cutoff frequency of the TE_{10} mode and ensure the same cutoff frequency for the TE_{20} mode. The bandwidth of these guides can be increased by a factor of 2.

3.2.2.2. *Circular guides*

The basic circular guide is represented in Figure 3.10. It does not have a (coaxial-type) central conductor. The following characteristics are identical to those or rectangular or square guides:

– it therefore cannot be treated in the same way as a conventional coaxial cable;

– it cannot deal with TEM waves, so transverse modes are generated by longitudinal modes;

– the modes are of the type $TE_{m,n}$ or $TM_{m,n}$;

– each mode has a cutoff frequency, below which it cannot propagate;

– each parameter in any mode (propagation constant, wavenumber, phase velocity, etc.) are all functions of the frequency (unlike the coaxial guide);

– the phase velocity is greater than the speed of light, unlike the group velocity.

These characteristics are common to all guides that do not have internal conductors.

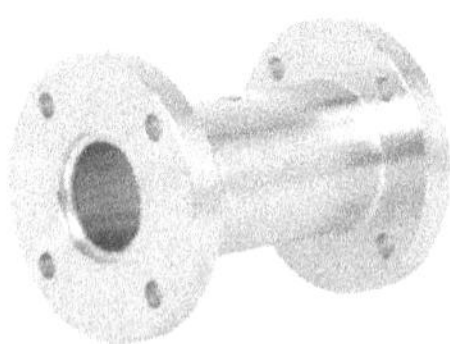

Figure 3.10. *Basic smooth cylindrical waveguide*

The cutoff frequencies involve the first root $\rho_{m,n}$ of the first-order Bessel functions for the TM modes and $\rho'_{m,n}$ for its first differential for the TE modes (Figure 3.11):

$$f_c(Hz) = \frac{\chi}{2.\pi.a}.v \qquad [3.5]$$

where:

– v is the velocity of propagation in the guide;

– a is the radius of the guide.

Table 3.1 shows the value of χ for the first six modes.

Mode	χ
TE11	1.841184
TM01	2.404826
TM11, TE01	3.831706
TM02	5.5020078
TE02	7.015587

Table 3.1. *Coefficients χ for the first six modes for a circular guide*

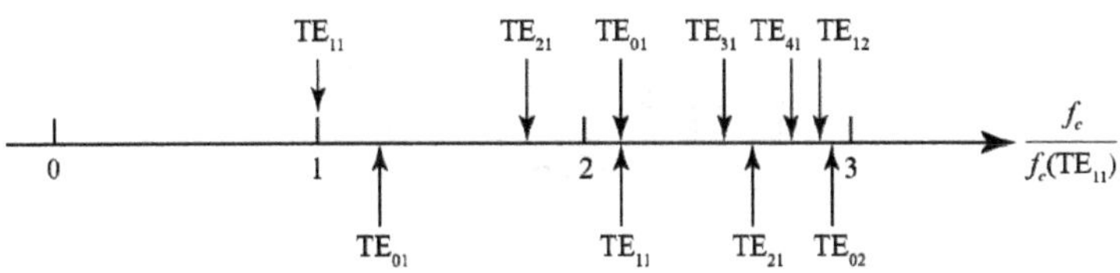

Figure 3.11. *Cutoff frequencies of the first ten modes for a circular guide*

The guided wavelength far from the cutoff frequency is given by:

$$\lambda_g = \frac{\lambda_0}{\sqrt{1-(\frac{\chi.\lambda_0}{2.\pi.a})^2}} \qquad [3.6]$$

where a = the radius of the guide.

The losses for the $TE_{m,n}$ modes are:

$$\alpha_{TEm,n} = \frac{R_S}{a\eta\sqrt{1-(\frac{f_{cTEm,n}}{f})^2}}[(\frac{f_{cm,n}}{f})^2 + \frac{n^2}{(\rho'_{m,n})^2 - n^2}] \text{ (Np/m)} \quad [3.7a]$$

The losses for the $TM_{m,n}$ modes are:

$$\alpha_{TMm,n} = \frac{R_S}{a.\eta\sqrt{1-(\frac{f_{cTMnm,n}}{f})^2}} \text{ (Np/m)} \quad [3.7b]$$

where $R_s = \sqrt{\frac{\pi.f.\mu_0}{\sigma}}$.

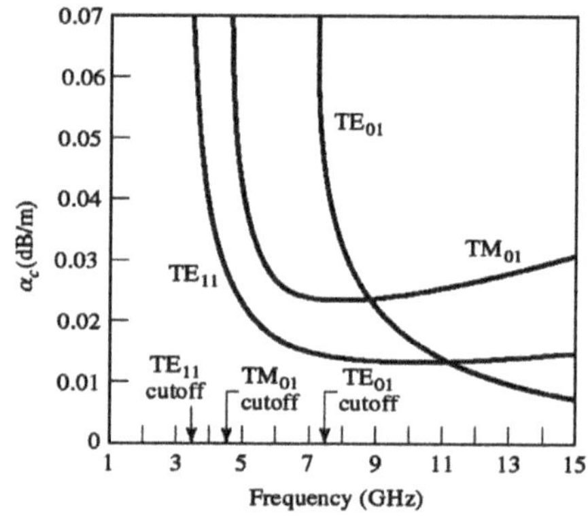

Figure 3.12. *Losses for different modes $TE_{m,n}$ and $TM_{m,n}$ for a guide 5.08 cm in diameter*

The excitation of a circular guide takes place in the same way as for rectangular or square guides, by means of coaxial probes or through slots, for the same reason of compatibility of the electric field lines of the fundamental mode TE_{11} with the field radiated by a probe or a slot.

As with TE_{10} mode, TE_{11} mode is the only one which propagates to the next modes. The odd-numbered modes TM_{01} and TE_{21} are used to produce satellite tracking signals. Given the compatibility of the TE_{10} and TE_{11} modes, there is no issue in switching from a rectangular or square guide to a circular one, using a progressive meridian transition in a guide, given the similarity of the electric field lines for the two fundamental modes.

3.2.2.3. *Ridged (corrugated) circular guide and hybrid mode*[3]

Corrugated waveguides and corrugated horns are very important where antennas with highly pure polarization and high light yield (> 85 %) are needed. Indeed, the diffraction disk or Airy disk in the focal plane of centered systems (similarly to in optics) has the structure of the hybrid mode HE_{11}. In addition, the radiation pattern exhibits perfect symmetry of revolution across a wide range of frequencies, which ensures maximum energy transfer (Figure 3.13a). in the focal plane of centered (i.e. focusing) systems, the fields have the local structure of a plane wave, which also allows us to use radius theory for their design.

For the smooth guide shown in Figure 3.13b, we see that the vector $\vec{E}$ of the TE_{11} mode has two components: a nominal component following an $\hat{x}$ axis or copolar axis, and an orthogonal component or crosspolar $\hat{y}$ axis, which the HE_{11} mode does not have.

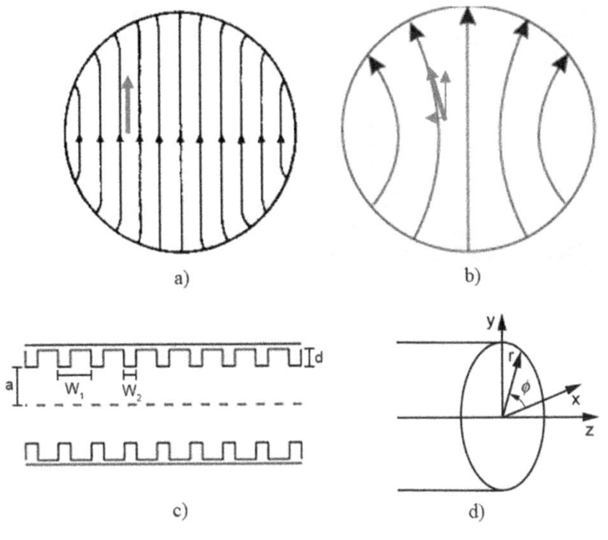

Figure 3.13. *Field line $\bar{E}$ for, respectively: a) hybrid mode HE_{11} for a corrugated guide or horn; b) TE_{11} mode for a smooth guide or horn; c) the cylindrical corrugated guide of radius a; and d) the system of axes of the cylinder*

3 (Drabowitch 1978; Clarricoats 1984; Huang 2008).

In this type of guide or horn, the behavior of the magnetic field in the plane E is the same as that of the electrical field in the plane H. Thus, we have weak radiated crosspolarization (see Figure 3.13a). In fact, we are dealing with an anisotropic surface. In order to make such a surface in favorable conditions, it is necessary to consider w_2 to be small with respect to w_1. The impedance expressed in relation to the surface of the corrugations is calculable by considering that each cavity constitutes a section of a line short-circuited at the bottom of the groove. The average impedance expressed in relation to the surface, therefore, is given by:

$$Z = j.\frac{w_1}{w_1 + w_2}.\sqrt{\frac{\mu_0}{\varepsilon_0}}.\tan\left(\frac{2.\pi}{\lambda_0}.d\right) = j.X$$

where λ_0 is the wavelength at the frequency f_0 for which the guide is being designed.

In order for the surface to be treated as a magnetic wall, we merely need to choose $d = \frac{\lambda_0}{4}$, which allows us to express the transverse admittance as infinite and longitudinal as null, over the surface. In fact, to break the current lines and decrease the value of the magnetic field, the impedance Z simply needs to be capacitive. This gives us the following condition:

$$\frac{\lambda_0}{4} < d < \frac{\lambda_0}{2}$$

This condition means that a wide range of wavelengths can be used for operation, making the device broadband. The bandwidths commonly used are 60%, or 100% in the case of more complex ridges.

The grooves must be thin and close together in relation to the wavelength to simulate a continuous anisotropic admittance. An important property of the hybrid modes TE and TM is that they have the same cutoff frequency and the same phase velocity.

The three main hybrid modes used are:

– the fundamental even-numbered mode HE_{11} used for "telecommunication" channels and illustrated by Figure 3.13a. Its amplitude at the aperture is a type-1 zero-order Bessel function: $E_x(r) \approx 0$, $E_y(r) = J_0\left(\mu_{01}.\frac{r}{a}\right)$, having the first root

$E_y(r=a)=0$ and $\mu_{01}=2.4$. This expression is the value E_0 given by equation [2.82];

– the two odd-numbered modes HE_{21} and TM_{01} used in the case of "monopulse" tracking have their amplitude at the aperture expressed by a type-1 first-order Bessel function.

The transition between a smooth guide and a corrugated (grooved) guide poses no particular problem because the field lines are compatible: the matching is made by taking a decreasing groove depth from around $\lambda_0/2$ to $\lambda_0/4$ at the junction.

Below, the applications and functional characteristics of grooved guides and horns are summarized:

– applications: high-yield antennas and high purity of crosspolarization;

– functional characteristics:

- very low ohmic losses, of the order of 0.01 dB/m;

- very broad frequency band in hybrid mode HE_{11};

- excellent radiation pattern independent of frequency in a wide range of frequencies, and great stability in the phase center;

- excellent power resistance (over 1 MW without a dielectric, and higher still in a vacuum);

- good mechanical stiffness;

- can be used for quasi-optical applications.

3.2.2.4. *The concept of characteristic impedance of a waveguide with any straight-edged profile*

For a coaxial guide, it is easy to define a characteristic impedance, as we do with conventional circuits, because there is a potential difference between the inner and outer conductors; in the same way, we can determine the direction of the currents flowing therein. The characteristic impedance is clearly defined by the voltage/current ratio $Z_c=\dfrac{V(z)}{I(z)}$, where z is the direction of propagation. In the case of guides without an internal conductor, it is more difficult to define a characteristic impedance. Indeed, the modes E and H do not derive from a scalar potential V, as do the TEM modes in the coaxial regime – a mode which does not exist in hollow waveguides. Therefore, we need to look directly at the fields, giving up the familiar language of voltages and currents. In a hollow guide, there is always a longitudinal component E_z and H_z that generates the transverse modes TE and TM. In a guide, it is impossible to clearly define the conventional concepts of voltage and current. This impossibility becomes evident when we think of a guide made of a simple metal tube, where we cannot distinguish two terminals to speak of a voltage between them. Nevertheless, efforts are made to invent characteristic impedances with the aim, above all, of introducing a figure to simplify the calculations, even if that value can only be set by agreement, and is devoid of the physical meaning that its name would seem to suggest. One popular idea is to take, as the characteristic impedance, the impedance η_E and η_H of the TE and TM waves, because such a definition has the advantage of being applicable to any waveguide and any mode. However, η_E and η_H represent a ratio between the transverse fields $\frac{E_T}{H_T}$, rather than a V/I ratio. We could make the electrical field E identical to V, and the magnetic field H identical to I, but this attempt yields the contradiction of failing to respect the power condition. In fact, *it must be borne in mind that the matching essentially means the absence of reflection*. The characteristic impedance, with a ratio between the transverse fields, therefore depends on the mode in question. We have the following impedances for the TE and TM modes in a guide filled with a dielectric, having the constants ε and μ and exhibiting no ohmic losses:

– for the TE modes:

$$Z_{mn} = \frac{j.\omega.\mu}{\gamma_{mn}}$$

– for the TM modes:

$$Z_{mn} = \frac{\gamma_{mn}}{j.\omega.\varepsilon}$$

with $\gamma_{mn} = \sqrt{(k_{mn})^2 - k^2}$, where k is the wavenumber in a vacuum, and $k_{mn} = \frac{2.\pi}{\lambda^c{}_{mn}}$ with $\lambda^c{}_{mn}$ being the cutoff frequency of the mode in question.

For a circular or rectangular guide:

$$Z_{TE} = \frac{Z_0}{\sqrt{1 - \frac{f_c^2}{f^2}}} \qquad [3.8a]$$

$$Z_{TM} = Z_0 . \sqrt{1 - \frac{f_c^2}{f^2}} \qquad [3.8b]$$

where:

– $Z_0 = \sqrt{\frac{\mu}{\varepsilon}}$, equal to $120.\pi \approx 377.\Omega$, the impedance of a vacuum where there is no dielectric;

– f, the working frequency;

– f_c, the cutoff frequency of the $TE_{m,n}$ or $TM_{m,n}$ mode in question.

By way of example, for a rectangular guide, the characteristic impedance for mode TE_{10} is:

$$Z_{TE10} = \frac{120.\pi}{\sqrt{1 - \frac{\lambda}{2.a}}}$$

which is a function of the frequency, depending only on the long side a of the guide, which is not the case with other modes (Silver 1984).

3.2.2.5. *Orthogonality of guided modes*

In a guide, the modes are orthogonal, so the projection of one onto the other should be zero. The projection is obtained by finding a scalar product, which needs to be integrated in the plane transversal to the guide's direction of propagation:

$$\iint_S \vec{E}_1 . \vec{E}_2^{\,*} ; dS = 0 \qquad [3.9]$$

This property is important because it allows us to extract the energy brought by a mode, by probe or slot, along the field lines and/or current lines. For example, the odd-numbered modes TM_{01}, TE_{21} or HE_{21} for satellite tracking are extracted using a specially studied mode extractor.

3.2.2.6. *Coupling and transfer function*[4]

The coupling between two fields $\vec{E}_1$ and $\vec{E}_2$ (and similarly for $\vec{H}_1$ and $\vec{H}_2$) is quite simply a normalized scalar product. Suppose that ($\vec{E}_1, \vec{H}_1$) is the field radiated by a source, immersed in an incident field ($\vec{E}_2, \vec{H}_2$). The power transfer coefficient from the incident wave to the source is given by:

$$\xi = \frac{\left|\iint_S \left(\vec{E}_1 \wedge \vec{H}_2 - \vec{E}_2 \wedge \vec{H}_1\right) ds\right|^2}{\iint_S (\vec{E}_1 \wedge \vec{H}_1^*) ds . \iint_S (\vec{E}_2 \wedge \vec{H}_2^*) ds} \qquad [3.10a]$$

where the surface S encloses the source. In the case of a horn, for example, S may be limited to the plane of the aperture. Perfect matching to the incident field corresponds to the condition $\vec{E}_2 = \vec{E}_1^*$, $\vec{H}_2 = -\vec{H}_1^*$, in which case we have $\xi = 1$. ξ represents the illumination yield factor of the antenna.

If ($\vec{E}_1, \vec{H}_1$) and ($\vec{E}_2, \vec{H}_2$) are plane waves, the expression [3.10a] can be reduced to:

$$\xi = \frac{\left|\iint_S \vec{E}_1 . \vec{E}_2^*\right|^2 .dS}{\iint_S \left|\vec{E}_1\right|^2 .dS . \iint_S \left|\vec{E}_2\right|^2 .dS} \qquad [3.10b]$$

This relation is important, because it allows us, for example, to quantify connection losses between two different guides. It also allows us to optimize a horn which radiates its fundamental mode, such as HE_{11} mode with the Airy disk in the focal plane of a centered system, such as reflector antennas. If the incident wave $(\vec{E}_2, \vec{H}_2)$ is a plane wave in any direction $\vec{u}$ toward the source, by calculating the transfer coefficient ξ, we find the source's directivity factor in that direction, and thus we can plot the source's radiation pattern.

4 (Drabowitch 1978; Picon 2009).

3.3. Supply to antennas, reference access, impedance matching and balun[5]

3.3.1. *Supply lines*

The supply lines are either waveguides or coaxial cables. Flexible or semi-rigid coaxial cables are permanently used in the context of so-called "low-frequency" applications, below 5 GHz – rarely higher, and even then, only temporarily. They are used where the transmitted power levels are between a few watts and a few kilowatts, as in the case of *indoor* or *outdoor* domestic applications, or ground mobiles with base transceiver stations (BTS). In the latter case, the distance between the radiating hardware and the HPA (*High-Power Amplifier*) or the LNA (*Low-Noise Amplifier*) is short, reduced to its minimum by technological means. The coaxial cables and connectors used, for these purposes, are of high quality, not only to ensure low ohmic losses, but to prevent the generation of passive intermodulation (PIM) products. The grounding, over a distance of ten meters or so, uses a lower-quality coaxial cable. For the highest frequencies, in the microwave range, coaxial cables are used after a high-to-low frequency transposer for downlink. The transposition is often made from L band after the LNA. For uplink, the reverse is true: the coaxial cables are used upstream of the high-frequency transposition upstream of the HPA, and then in guides to the antenna, depending on the power of the HPA and the antenna technology used.

3.3.2. *Reference access*

The electrical and radioelectrical parameters (VSWR, ohmic losses, gain, antenna yield, decoupling between access points) of an antenna are always calculated and measured in a reference plane which defines the position of the access points: for uplink – i.e. transmission – after the high-power amplifier; for downlink – i.e. reception – before the low-noise amplifier. In the case of frequency reuse by orthogonal polarizations, the pathways are doubled without necessarily having the same plan of reference. In the case of tracking with a mode extractor, we need to define a plane of reference for it. These different planes contractually define access in a set of technical specifications.

Figures 3.14a and 3.14b illustrate these accesses and the plane of reference where the total losses $\Sigma loss$ are taken into account. These losses are inherent to the technology used to manufacture the antenna. Once the antenna has been adapted to

---

5 (Silver 1984).

its supply, the ohmic losses are the primary cause of degradation of its yield and its internal noise temperature:

$$\Gamma = \frac{(Z_{ant} - Z_c)}{(Z_{ant} + Z_c)}$$

where $0 \leq |\Gamma| \leq 1$ is the reflection coefficient at the *transmission* or *reception* access point, with Z_{ant} being the impedance of the antenna on the antenna side, and Z_c the impedance of the supply cable on the side of the HPA or LNA.

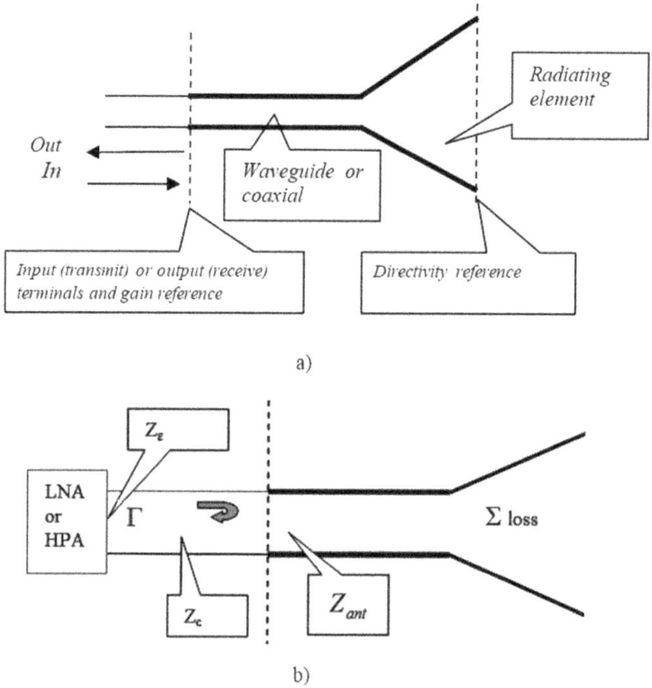

Figure 3.14. *a) Reference access for "transmission" or "reception"; b) Reference access and losses in an antenna*

The VSWR (*Voltage Standing Wave Ratio*) is given by:

$$vswr = \frac{(1+|\Gamma|)}{(1-|\Gamma|)}, \quad 1 \leq vswr \leq \infty$$

From the standpoint of a *system*, the requirements differ depending on whether it is transmitting or receiving. When transmitting, the most important thing is to transmit sufficient power in the desired directions; therefore, the radiation efficiency and the gain of the antenna are important parameters in a link budget. When receiving, a crucial requirement is typically that the antenna has a good carrier-to-noise ratio (C/N), which is connected to the radiation efficiency, the gain of the antenna and the transfer of as much of the energy received to the receiver as possible.

3.3.3. *Matching networks*

The problem of impedance matching is often encountered in the case of wired technologies, of printed antennas and miniaturization. Matching is not always possible, as is the case with antennas with very low radiation resistance R_r, such as antennas which are much smaller than the wavelength, like the long-wave and *medium-wave* antennas used by onboard radios in cars, whose radiation resistance is only a few milliohms. Such is also the case with mobile phone antennas, where their impedance varies depending on the position of the user's head and hands. Matching, therefore, is average compensation, facilitating satisfactory operation. For the other technologies such as "aperture" antennas, the impedance for the transitions is normalized at 50 Ω, and all other hyperfrequency components such as couplers, circulators, loads, etc.

Impedance matching entails eliminating the so-called *mismatch loss*, so cancelling out the reflection coefficient Γ by transforming the impedance with a matching network. To produce such a network, we use either discrete passive components or transmission cables (which can be viewed as a network of distributed passive components), as illustrated by Figures 3.15 and 3.16.

Matching circuits or networks are exclusively inductors and capacitors. They are often used in printed technology in the microwave range. For low and very low frequencies, in view of the large wavelengths, the bulk and the power that may need to be transmitted, the networks are classic LC components based on coils for self-inductors and capacitors. The matching networks are exclusively LC circuits of greater or lesser complexity.

We might be tempted to introduce serial or parallel resistors to bring the real part of the impedance to Z_c. Although this matches the load, it does not improve the active electrical power supplied to the antenna. Indeed, the active power delivered to the matched load will be dissipated as heat from the resistor we have added, and will not help to improve the antenna's radiation. The drawback to these matching devices is that they are narrow-band because, by their principle, they are produced at the frequency f_0 at the center of the bandwidth $B = \frac{f_{max} - f_{min}}{f_0}$.

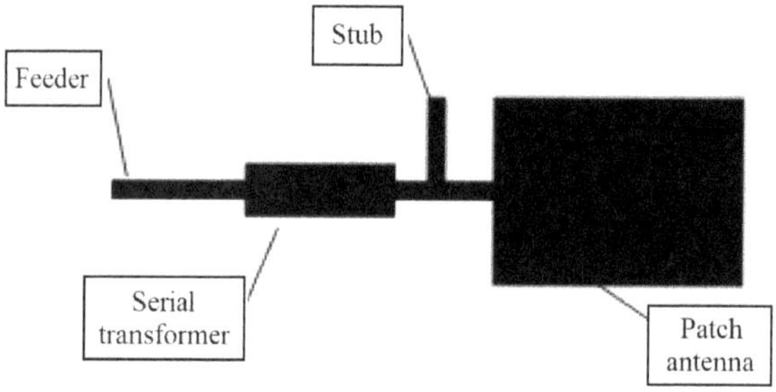

Figure 3.15. *Principle of a matching device*

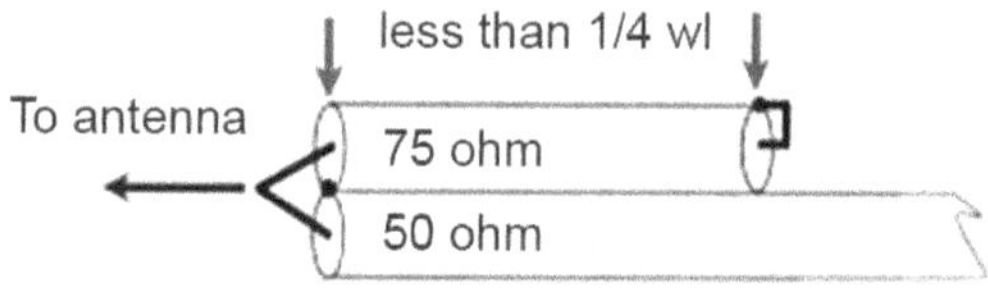

Figure 3.16. *Matching device for a cable antenna*

A technique used for a wire-form antenna (and especially *Yagi-Uda* antennas) is to take a *T-match* or a *Delta-match*, or even a *Gamma-match*.

Note, for example, that to match a dipole or a monopole, we can act on the cross-section of the strands, the effect of which is to decrease the quality factor Q, by lowering the input impedance and increasing a fraction of ohmic losses (Balanis 2008; Huang 2008).

For waveguides, the goal is to match the impedance of the fundamental mode TE_{10} for rectangular guides, or TE_{11} for circular guides, to the impedance of the coaxial transition, which is 50 Ω. There are a variety of ways in which this can be done. Firstly, we can use an iris, as shown in Figures 3.3 and 3.17. Depending on the irises or screws, the effect will be inductive or capacitive (Silver 1984). By combining the arrangement of the irises as in Figure 3.17, we obtain a self-inductance coil and a parallel capacitor, which, in certain cases, increases the bandwidth. Impedance matching can also be done by altering the dimensions on the guide. Note that, when the profile is altered towards its mouth, a horn behaves like a progressive impedance transformer, transforming the impedance from 377 Ω (its value in a vacuum) to the impedance of the guide, and then 50 Ω at the transition from the guide to the coaxial cable (see Figure 3.3).

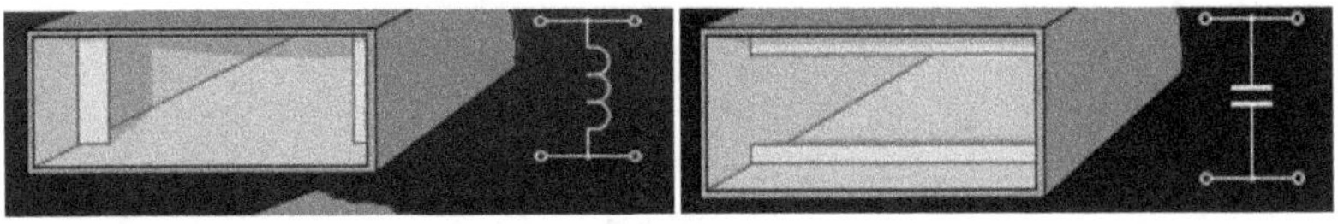

Figure 3.17. *Inductive and capacitive effects depending on the position of irises within the guide*

3.3.4. *Baluns and symmetrizers*

A balun is not a matching device, as discussed above, instead, it is a device which transfers impedance from one antenna to another: the power supply antenna. It does not improve the vswr.

There are two types of balun: voltage symmetrizer baluns and current symmetrizer baluns. With the former, the output means that we can have two voltages that are always equal, whatever the impedances (to a certain degree) of the antenna and of the power supply. For the latter, the output of the balun or of the symmetrization device means that we can have two currents that are equal, whatever the values of the impedance (to a certain degree), and here the voltages may be different if the impedances are. This latter type of setup is used to switch from a coaxial to a symmetric system, to prevent an external current on the coaxial cable which will interfere with the radiation pattern. It operates on the principle of the shock self-inductance coil to eliminate the HF current on the outer conductor of the coaxial cable (Figure 3.18). Note that a circuit called a BALUN (*balance/unbalance*) is a symmetrizer and is a particular example of an impedance-matching transformer. It is known as a symmetrizer because it renders symmetrical the radiation pattern of a dipole, for example, by eliminating the current on the external shield of the coaxial. A balun may be similar to a voltage transformer, comprising a primary coil and a secondary coil, wrapped around a straight spool, or a torus made of ferrite or another material. From the number of spires of the primary and the number of spires of the secondary, we can determine the ratio of conversion from the impedance of the antenna to that of the coaxial supply cable. For example, a ratio of 1/1 means that the impedance stays the same; a ratio of 2/1 doubles the impedance, meaning we can switch from 100 Ω to 50 Ω; a ratio of 4/1 allows us to switch from 200 Ω to 50 Ω; and so forth. A balun does not enable monoband antennas to operate in multiband mode. In summary, a balun has only two purposes: to insulate the transmission line and the load, and to balance the output currents and switch from one impedance to another in a given ratio.

Figure 3.18 shows an example of a balun applied as a symmetrizer to a dipole.

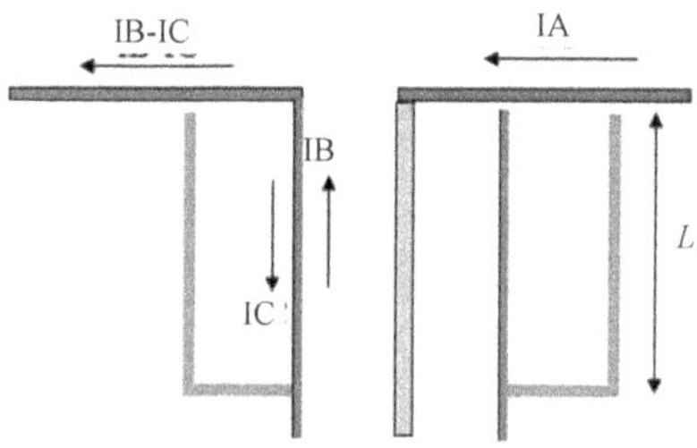

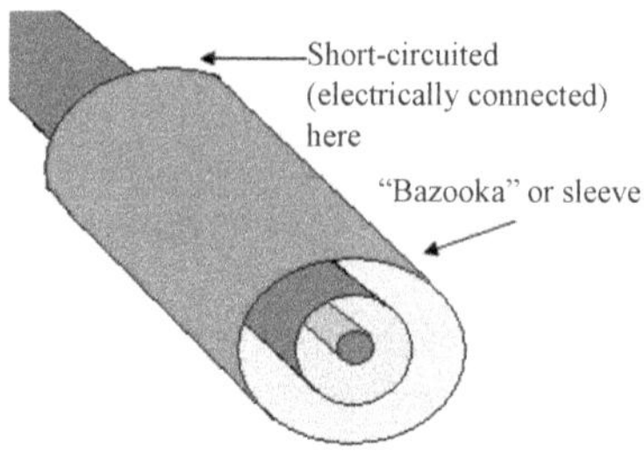

Figure 3.18. *For a dipole: a) operation of the sleeve balun; b) bazooka or sleeve balun*

The bazooka consists of a short-circuit (length $L = \lambda/4$) over the external shield of the coaxial feeder.

3.4. Reflector antennas[6]

Reflector antennas are employed for links which require a high antenna gain (gain in dB = directivity in dB – losses in dB) in one or more specific directions with their radiation patterns as spatial filters, which help to combat frequency jamming and manage coverage over a given geographical region. This technology, which has gained in maturity over a very long period of time, is built into geostationary

6 (Drabowitch 1978; Silver 1984; Milligan 2005; Balanis 2008).

satellites, primarily, which can provide fixed or dynamic management of terrestrial coverage, as shown in Figure 3.19. In 1993, the advanced communications technology satellite (ACTS) was a NASA prototype, which took its first steps into advanced antenna technology and communications, giving the world the first successful commercial use of the spectrum in the Ka band.

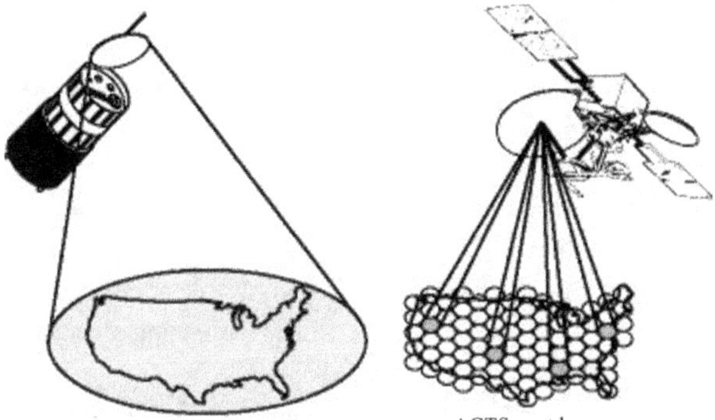

Figure 3.19. *Terrestrial coverage with onboard reflector antennas on satellites: global coverage satellite (1970) and multimission (1993)*

Gradually, printed technology is coming to replace reflector antennas, more specifically for scrolling satellite constellations in low, medium or high orbit, because they are less costly and less bulky for the useable load and they offer more extensive possibilities for dynamic beam management.

The diameter D of reflector antennas, depending on the ratio D/λ, ranges from a few tens of centimeters in Ka-band equipment to tens of meters in the L, C and Ku bands (Figures 2.8b and 3.20). These antennas, which operate at high frequencies, are described as quasi-optical antennas, in that the reflecting surfaces are larger than the wavelength. In this case, an incident EM plane wave is reflected in the geometrical optical sense, at any given point P_i. For a centered system, due to summation of the *rays*, it exhibits the structure of a plane wave in the focal plane. At any point P_i on the reflectors, the currents are given by the physical optics (PO) relation $\vec{J}_i = 2.\left(\hat{n} \wedge \vec{H}_{incident}\right)$, and do not depend on the neighboring points, but instead, only on the magnetic field $\vec{H}_{incident}$ associated with a ray. Next, all the laws of geometrical optics apply. When we know the currents $\vec{J}_i$ at all points on the surfaces, the total EM field at any given observation point, be it near or far, is calculated using the EFIE or MFIE equation given by equations [2.77a] and [2.77b], where the volume integral is replaced by a surface integral (Silver 1984). By surface sampling, the double integral $\iint$ is replaced by $\sum$ the sum of the elementary surfaces ΔS_i across which the current $\vec{J}_i$ is supposed to be constant, because it has a slow variation of amplitude and phase in relation to the wavelength λ. These samples are often called *facets*. The use of geometrical optics (GO) and physical optics (PO) means that the EM coupling of the excitation source with the reflectors is negligible, as is the coupling between reflectors. These asymptotic methods are powerful tools which do not require much computation power to develop this type of antenna. In the case of compact antennas, where the coupling between the different components is non-negligible, the EFIE or MFIE equations are solved numerically or analytically (Figure 3.20), for example, with the SRSR software, developed by Orange Lab (symmetrical revolution radiating structures) based on the rigorous solution of the integral equations.

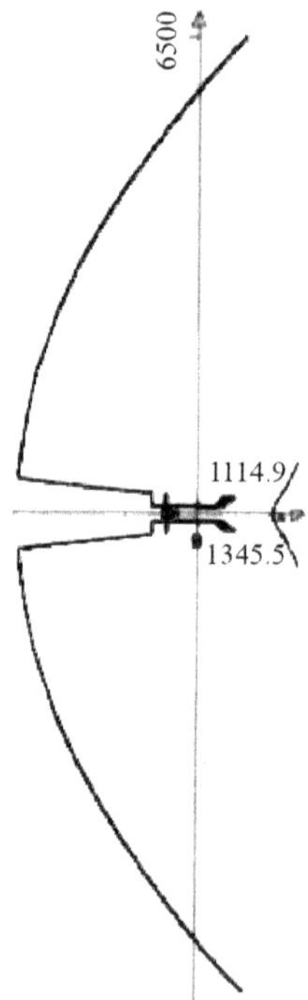

Figure 3.20a. *Cassegrain antenna 13 m in diameter, illuminated by a corrugated horn*

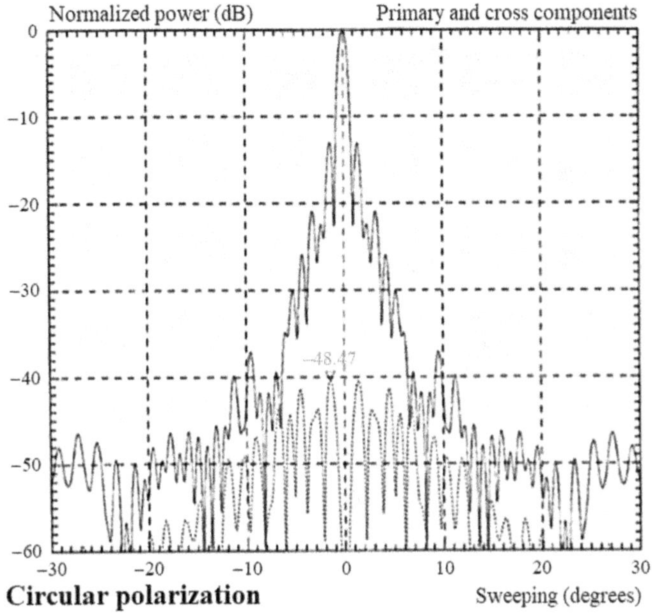

Figure 3.20b. *Radiation patterns in nominal polarization (copolarization) and crosspolarization at 1.5 GHz for a directivity of 46 dBi*

The reflectors in the antenna shown in Figure 3.20 are "shaped", making use of geometrical optics. The profile of the two reflectors is slightly deformed, so that, at the aperture of the large reflector, we obtain as constant a distribution of the EM field as possible, in terms of both amplitude and f phase, so as to optimize the illumination yield up to 85+% (aside from mechanical imperfections in the manufacture), compared to the theoretical maximum reference directivity of 100%. This reference directivity is obtained with a circular aperture that would capture all the energy of an incident plane wave from infinity. For reasons of manufacturing costs, for intermediary diameters, only the small reflector is shaped. Note that at a constant diameter, the higher the directivity of the antenna, the lower are the diffuse lobes: here − 50 dB at 20° of displacement. In the case of so-called *centered* antennas, the source and the secondary reflector exhibit a mask over the aperture of the large reflector, which limits the illumination yield. This problem does not occur with *offset* antennas, meaning they have a displaced source and secondary reflector. Therefore, a comparable yield can be achieved without shaping.

For small antennas where there is a not-insignificant degree of coupling between the different constitutive elements, a method employing a numerical solution of the EFIE and/or MFIE integral equations is crucially important. For example, the SBFA (*Short Backfire Antenna*), whose diameter is 2λ, shown in Figure 3.21, is designed with SRSR.

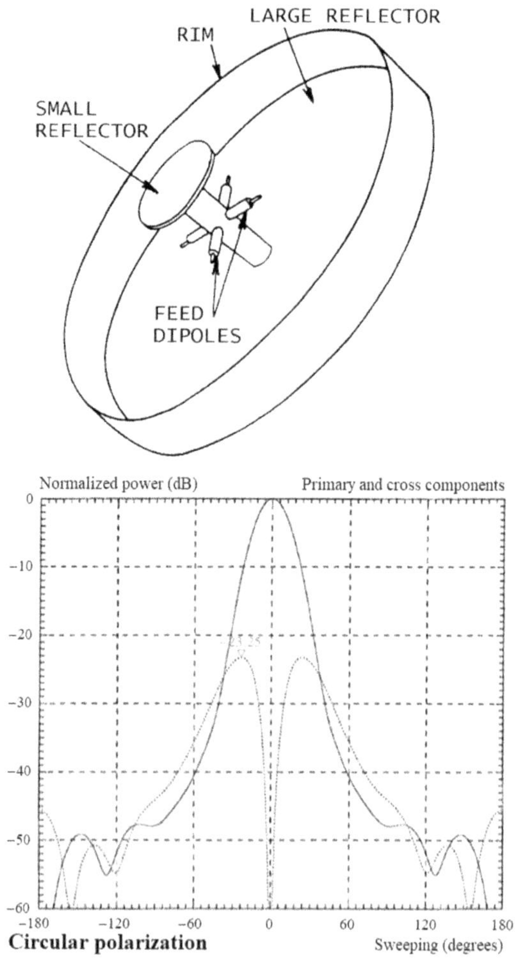

Figure 3.21. *Radiation pattern of an SBFA antenna with diameter 2λ at 1.5 GHz in circular polarization for a directivity of 17.7 dBi*

3.5. Printed antennas[7]

Wireless, spatial communications and radar often use reflector antennas. They are high-performing devices that produce a good yield, highly pure polarization and a large bandwidth in terms of frequency. In view of slight ohmic losses, they can deal with power levels up to several megawatts, as with long-range radars. In the case of mobile applications for new-generation satellites, their weight, their bulk, their real-time reconfiguration and their cost are major drawbacks. For these applications, printed technology offers a number of significant advantages as numerous reasons to choose them, from the classic patch (Figure 3.22) with small bandwidth to spirals and other Vivaldi antennas, which have very large bandwidth. For a detailed discussion, readers can consult the subject-specific literature – in particular, James (1989).

The benefits are:

– low manufacturing costs;

– the antennas are small and light;

– the possibility of printing them on non-flat surfaces, and even on flexible surfaces;

– the possibility of networking them in order to increase directivity, and the possibility of adaptive sweeping;

– the possibilities of integration into mobile devices;

– the possibility of achieving linear or circular polarization by adjusting the geometry of the patches, or agile polarization by electronic switching.

However, there are as many drawbacks as there are advantages, which limit the applications of printed technology:

– low gain, of around 5 dBi in single layer and 9 dBi in multilayer;

– difficulty in obtaining pure polarization across a wide range of frequencies;

– the possible excitation of surface waves in the dielectric. This can be controlled by strategies that render the structure more complex (e.g. the inclusion of pins or BIP (bandgap material) at the surface, etc.);

7 (James 1989; Milligan 2005; Balanis 2008).

– low levels of transported power in comparison to other antenna technologies, in view of the various types of dielectric and metal losses in the various circuits that make these antennas up;

– low bandwidth, for equivalent performances, in comparison to reflectors.

3.5.1. *Low-bandwidth structures*

Low-bandwidth structures are based on resonances, as are dipoles and other wire monopole antennas, or on cavities with apertures such as slots. The more open the cavities are, the wider will be the range of frequencies. An example of a slot cavity can be seen in printed antennas such as patches (Figure 3.22).

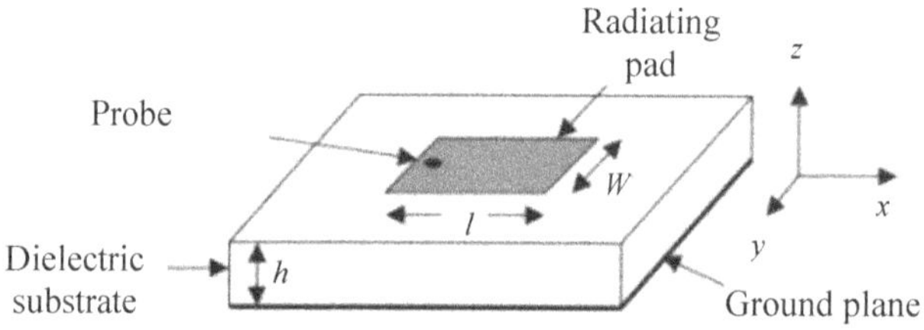

Figure 3.22. *Basic structure of a printed antenna*

We use the terms *printed antenna*, *plate antenna* and *patch antenna* interchangeably, to speak of the flat antennas made by photoengraving. The basic printed antenna (Figure 3.22) consists of a fine metal film (the patch) over a ground plane, both of which are engraved on both sides with a substrate whose permittivity is ε_r and thickness h, such that $0.003.\lambda_0 < h < 0.05.\lambda_0$. There are planar antennas that are not made by photoengraving, such as PIFAs (*Planar Inverted-F Antennas*), which are made with a PEC (copper, aluminum), of a certain thickness, and short-circuited halfway along the length of the patch (Figure 3.23). In this case, the dielectric is replaced with air, allowing for easier integration into handsets, for example.

There is similarity in the way of working between this family of printed antennas and wire antennas: the distributions of currents which generate the radiated field are imposed by *resonating modes*. In the case of wire antennas, these are one-dimensional steady-state modes; in the case of printed antennas, they are three-dimensional steady-state modes. The nature of the modes depends on the shape of the patch. As is the case for practically all types of antennas, whether wire, aperture or printed, an analytical solution is possible in a limited number of cases with simple basic structures, using a printed pattern in the form of a rectangle, a circle, a ring, etc. These analytical formulations allow us, as with other types of antennas, to pre-size the antenna using rigorous numerical simulations.

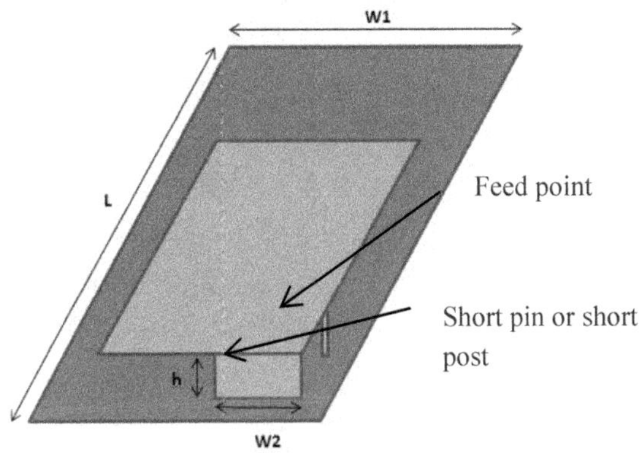

Figure 3.23. *Basic metal PIFA*

A printed antenna can be viewed as an open resonating cavity, composed of four magnetic lateral walls and two electrical horizontal walls. The radiation is produced by the leakage of the field at the extremities between the metal patch itself and the ground plane (see Figure 3.24). The operation of the antenna is therefore illustrated by two equivalent radiating slots at the edges and separated by the length L. For maximum power transfer to the antenna, the length L (Figure 3.22) is chosen to be slightly less than the guided half-wavelength λ_g in the dielectric. This operation corresponds to the excitation of the fundamental mode (fundamental resonance frequency), which corresponds to the lowest excited frequency.

In the case of a rectangular antenna with the dimensions L and W, the resonance frequencies of a TM mode $TM_{m,n}$ in the cavity are given by:

$$f_{mn} = \frac{c}{2.\pi.\sqrt{\varepsilon_r}}.\sqrt{\left(\frac{m}{L}\right)^2 + \left(\frac{n}{W}\right)^2} \qquad [3.11]$$

which is comparable to equation [3.3] for rectangular-shaped guides. The fundamental mode here is the $TM_{1,0}$ mode. For a circular patch, it would be the TM_{11} mode.

The bandwidth Δf (equation [3.12]) is inversely proportional to the quality factor Q of the antenna given by equation [3.13] which includes the square root of the effective dielectric constant ε_{effec} of the substrate and the height of the patch away from the ground plane. It takes account of matching with the voltage standing wave ratio (VSWR):

$$\frac{\Delta f}{f_0} = \frac{VSWR - 1}{Q.\sqrt{VSWR}} \text{ (in \%)} \qquad [3.12]$$

where f_0 is the resonance frequency of the antenna and Q the global quality factor – that is, including the quality factor due to radiation losses, conduction losses in the metal, in the dielectric, and surface waves (equation [3.21]).

For a rectangular patch, the quality factor is given approximately by:

$$Q = \frac{c.\sqrt{\varepsilon_{effec}}}{4.f_0.h} \qquad [3.13]$$

Formula [3.12] indicates that, in order to increase the frequency bandwidth, we need to take action on the quality factor Q and decrease it. This is tantamount to increasing the volume of the cavity by increasing h and/or decreasing ε (equation [3.13]). To increase the volume, an effective solution is to multiply the layers, as shown in Figure 3.25, which enable us, depending on one or more specified reference parameters – the vswr, the polarization, the directivity, etc. – to achieve over 30 % bandwidth by using dielectrics with a constant close to $\varepsilon \approx 1$; this value can be attained with certain low-density machinable foams. We can also use patches of different dimensions. However, there is a not-insignificant risk of surface waves appearing.

The fundamental equations for the design of a rectangular patch (Figure 3.22) are:

– for the effective dielectric constant ε_{eff}:

$$\varepsilon_{reff} = \frac{\varepsilon_r + 1}{2} + \frac{\varepsilon_r - 1}{2}\left[1 + 12\frac{h}{W}\right]^{-0.5} \qquad [3.14]$$

where:

- ε_{eff}, the effective relative permittivity;
- ε_r, the relative permittivity of the dielectric;
- h, the thickness of the dielectric;
- W, the width of the patch;

– *for the length L of the patch*: the patch radiates as though through two slots whose dimensions are $W \times \Delta L$, where:

$$\Delta L = 0.412.\frac{\left[\left(\varepsilon_{reff} + 0.3\right)\left(\left(\frac{W}{h}\right) + 0.264\right)\right]}{\left[\left(\varepsilon_{reff} - 0.258\right)\left(\left(\frac{W}{h}\right) + 0.8\right)\right]} \quad [3.15]$$

The length L of the patch is given by:

$$L = \frac{\lambda_0}{2} - 2.\Delta L \quad [3.16]$$

The effective length of the patch is:

$$L_{eff} = \frac{c}{2.f_0\sqrt{\varepsilon_{reff}}} \quad [3.17]$$

The width W is given by:

$$W = \frac{c}{2.f_0\sqrt{\frac{\varepsilon_r + 1}{2}}} \quad [3.18]$$

The expression of the impedance depends on the supply method, using a coaxial cable or microstrips (James 1989).

For an in-depth study of such planar technology, readers may usefully refer to Volumes 1 and 2 of J.R. James and P.S. Hall, *Handbook of Microstrip Antennas* (James 1989).

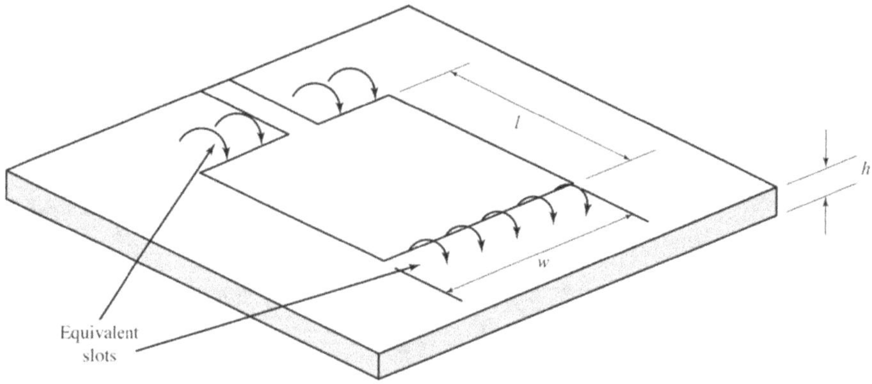

Figure 3.24. *Radiation of the patch represented by two slots of width ΔL*

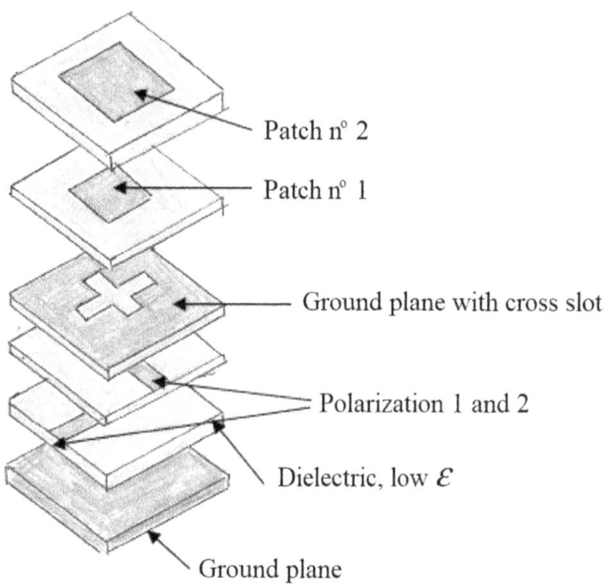

Figure 3.25. *Wide-band structure (≈ 30%), multilayer with double polarization*

On the basis of the fundamental versions of patches or PIFAs, complex structures have branched off, achieving impressive miniaturization for multi-band and wide-band applications (section 3.10).

3.5.2. *High-bandwidth structures, or frequency-independent structures*[8]

The definition of *broadband* is an arbitrary concept, depending on the peculiarities of the antenna. A practical definition given as a reference is as follows: if the impedance and the radiation pattern do not change significantly over an octave $\left(f_{high} / f_{low} = 2\right)$, then the antenna is described as broadband or very broadband.

In the case of printed antennas, to eliminate the cavity effect, we eliminate the ground plane and leave only the metal pattern engraved on the other face on the dielectric substrate. In order to be broadband, an antenna requires a physical structure whose geometry changes gradually, rather than sharply, as in a Vivaldi antenna (Figure 3.26a), where no area of the antenna exhibits discontinuities that could create a localized diffraction zone or a resonance zone, depending on the frequency. The phenomena of coupling also extend the bandwidth for certain resonating antennas: coupling takes place either in regard to the excitation, or in regard to the shape of the antenna, which introduces an extra resonating element, as shown in Figure 3.25, for example. As with wire antennas, they are fed by a coaxial cable or printed technology (stripline, coplanar, microstrip, etc.), with the same drawbacks in terms of symmetrization.

For frequency-independent antennas, in 1954, the work of a number of academics – mainly Rumsey and Du Hamel – began enabling engineers to retain a broadband of frequencies without drastically reducing the radiation pattern based on the principle of homothety. The linearity of Maxwell's equations gives rise to a principle of similarity which can be summarized thus: if we reduce the linear dimensions of a radiating system by a given ratio, the electromagnetic properties will remain the same, provided the wavelength is reduced by the same ratio. For a radiating structure defined exclusively by angles (for example, a cone), the antenna can be treated as identical to its homothetic reduction. Hence, at any and all frequencies, its electrical properties should remain unchanged. However, the antenna can be defined solely by angles only if its limited dimensions do not cause an extremity effect, setting the original structure apart from its homothetic reduction. This is not the case with a conical antenna, in which very significant extremity effects occur. Therefore (Drabowitch 1978), two conditions are needed in order for an antenna to be frequency-independent:

– it must be able to be superposed upon its homothetic reduction (a purely geometric notion);

8 (Rumsey 1957; Du Hamel 1957; Drabowitch 1978; Balanis 2008).

– it must have no extremity effect (an electromagnetic concept, which is not immediately apparent, and depends on the current distribution). They are limited at low frequencies by their finite dimensions, meaning that an extremity effect inevitably manifests itself beyond a certain wavelength. They are limited at high frequencies by the precision of their manufacture.

There are two types of frequency-independent antennas: "logarithmic periodicity" antennas, whose features are periodic functions of the frequency logarithm, and "logarithmic spiral" antennas. For an in-depth study, readers may usefully refer, for example, to Balanis (2008).

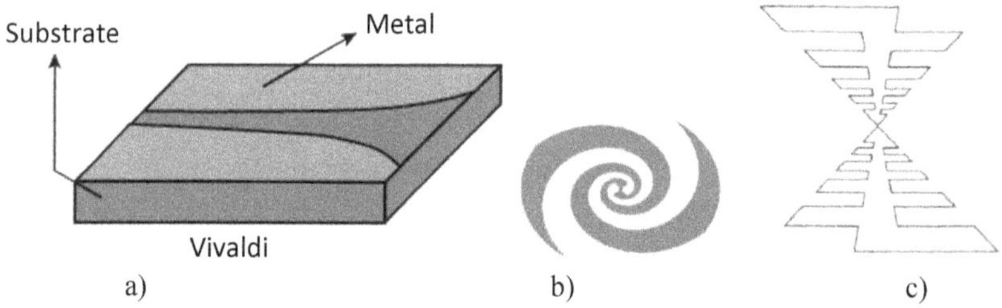

Figure 3.26. *a) Broadband Vivaldi antenna; b) and c) frequency-independent antennas: logarithmic spiral antenna and log periodic antenna*

3.6. Reference wire antennas[9]

Dipoles, monopoles, loop antennas, biconical dipoles and other antennas with narrow bandwidth or a very wide range of frequencies – such as spiral antennas, log-periodic antennas, etc. – and the associated networks are the most commonly used antennas in VHF and UHF bands, in the fields of radio communication, telecommunications, measurement of electrical and magnetic fields (Figures 1.3 and 1.8). For analytical calculations pertaining to these antennas and their applications, refer to the plentiful literature on the subject.

As with all antenna technologies, for a reliable result taking account of the most realistic geometric and dielectric data as possible, the analytical formulations are merely approximations, which can be used for a preliminary design, particularly for resonating antennas with narrow bandwidth, where the manufacturing tolerance margins are drastic. We therefore use computational codes employing a n umerical method of solving Maxwell's equations, on the basis of an integral or local formula, such as the

9 (Silver 1989; Tapan 2000; Milligan 2005; Taflove 2005; Balanis 2008).

method of moments (MoM), the finite element method (FEM) or the finite-difference time-domain method (FDTD), etc. These codes may take account of the geometric tolerance margins in the manufacture. The most "popular" method is FDTD because it allows us to visualize the currents as a function of the timestep, as well as the radiated EM field. In addition, computer-assisted design (CAD) codes can be used to display the development of antennas on screen. To find out more about this, readers are invited to consult the abundant literature on the subject, including the references quoted above and listed at the end of this book, as well as the recommended further reading.

3.7. Quality factor and frequency bandwidth[10]

3.7.1. *Quality factor*

Numerous antenna technologies are based on creating resonance with one or more radiating components at one or more frequencies. Thus, they can be approximated by an RLC electrical circuit, of greater or lesser complexity depending on the case. Like these circuits, they are characterized by a quality factor Q and a frequency bandwidth Δf on the resonance curve, in general at -3 dB from the field value. The field value is an important parameter for any component and electrical or electronic system. The quality factor Q measures the amount of total energy stored, with respect to the average power lost per cycle. The bandwidth of the antenna is maximized when the dissipated power is high. In other words, a low quality factor Q is necessary to obtain high bandwidth. The quantity of energy radiated, rather than the ohmic or dielectric losses, determines the efficiency of the antenna. Unlike an inductive or capacitive circuit with low losses, which requires a high quality factor Q, an antenna requires a low quality factor. The quality factor Q is given by:

$$Q = \frac{(total.energy.stored)}{(average.load.loss)} = \omega.\frac{W_E + W_M}{P_L} \qquad [3.19]$$

where W_E is the energy stored by the electrical field, W_M is that stored by the magnetic field and P_L is the average power delivered to the antenna.

At the resonance, the energy of the magnetic field and that of the electrical field have the same amplitude. The quality factor Q_0 becomes simpler, and is given by:

$$Q_0 = \frac{2.\omega_0.W_E}{P_L} = \frac{2.\omega_0.W_M}{P_L}$$

10 (Huang 2008).

When we know the quality factor Q_0 of a resonating circuit – here an antenna – we are able to determine the bandwidth $B = f_2 - f_1$ at mid-power (or − 3 dB):

$$Q_0 = \frac{f_0}{B} \tag{3.20}$$

where f_0 is the resonance frequency.

The bandwidth of an antenna corresponds to the frequency band where the energy transfer for power feed to the antenna or the receiver is greatest. Within the bandwidth, the reflection coefficient is low.

The quality factor Q tends toward infinity in the lossless case $R = 0$ in a circuit. For an antenna, because of radiation, Q falls. It will fall further by combining the losses in the dielectrics and in the conductors of which the antenna is made. Thus, we can deduce the effectiveness of radiation η_r by the ratio between the final quality factor Q and that due only to radiation losses:

$$\eta_r = \frac{Q}{Q_r}$$

The relation between the quality factors for each of the types of losses is written as follows:

$$\frac{1}{Q} = \frac{1}{Q_r} + \frac{1}{Q_d} + \frac{1}{Q_c} \tag{3.21}$$

where Q_d is linked to losses in the dielectric and Q_c to the conduction losses.

The factor Q is established by means of the ratio between the stored energy and the average energy (dissipated per cycle). This latter energy value is also linked to the average dissipated power $P_{dis} = P_L$.

3.7.2. *Frequency bandwidth*

The frequency band within which a system is required to operate is generally specified as a range (for single-band applications) or several ranges (for multiple bands) of frequencies where the antenna must satisfy, for example, a certain energy loss for a given reflection coefficient, a level of cross-polarization in the radiation pattern, or another criterion of the designer's choice. Ultimately, any designer may choose the criterion/criteria of interest to them. Thus, the concept of *narrowband* and *broadband* is fuzzy and ill-defined, but it is crucial to reach agreement on what is needed in technical exchanges. An antenna is called to function from a frequency f_{min} to a frequency f_{max}. One of the accepted definitions of bandwidth is as follows:

$$\%Bandwidth = 200.\frac{(f_{max} - f_{min})}{(f_{max} + f_{min})} \quad [3.22]$$

If we adopt this definition, a bandwidth below 40% is considered narrowband, whereas anything above that threshold is considered broadband. An antenna whose impedance and radiation pattern are stable over an octave and more ($\frac{f_{max}}{f_{min}} \geq 2$) is considered to be very broadband, or frequency-independent.

3.8. Miniaturization[11]

Electrically short antennas, meaning ones that are short in comparison to the wavelengths which they use, have, for many years, been the subject of numerous development efforts. Today, one of the most significant challenges facing designers is to allow miniaturized electronic devices to be integrated into communicating things (in the sense of the Internet of Things), the dimensions of which are intended to be ever smaller. The proliferation of communication standards is merely increasing the need for integration and miniaturization because in today's world, multiple antennas may need to fit on the same mobile terminal, serving different functions. Thus, the antenna is situated in a relatively complex environment, which will inevitably affect its performance. Antennas' radioelectrical performance is subject to the immutable laws of physics, and therefore it is possible to have antennas almost however small we want them to be. However, the crucial issue is that of the radio *link budget*, because we cannot indefinitely reduce the size of an antenna without compromising the quality factor, the bandwidth and the radiation pattern.

The frequency range to which these imperatives relate is mainly the UHF range from 900 MHz to 6 GHz, approximately, for mobile telephony (GSM), UMTS, DECT, wi-fi, Bluetooth and RFID. Sometimes, there are also esthetic or ergonomic

11 (Wheeler 1947; Chu 1948; Stutzman 1981; Felber 2000; Sarychev 2007; Huang 2008; Pozar 2009; Grange 2010).

concerns that must be dealt with which further complicate antenna design. In effect, the antenna is the most bulky component in a system, and it is no easy task to *miniaturize* it while preserving good radioelectrical performance. The higher the frequency, the smaller the antennas become, physically, in relation to the wavelength. But in the process, other technological problems arise, such as the feasibility of making circuits while retaining control – amongst other things – of engravings to limit unwanted radiation, or ohmic losses, for example. Current solutions for miniaturization – or, more accurately put, for *reducing antenna size* – are based on a compromise between the bandwidth, size and effectiveness of radiation, the imperative being to achieve a certain radio link budget which may be more or less strict, depending on the application. The materials used are conventional LHI materials. As they have reached maturity, current miniaturization techniques have led research in new directions based on new materials that exhibit different properties (both electrical and mechanical). Natural magnetic materials with high permeability and low losses are offering up new potential solutions. Similarly, materials with artificially created properties, which do not occur naturally – often called *structured composite materials or metamaterials* – are shifting the paradigm, thanks, in particular, to the progress in microelectronics, which means they can be made. These materials generally take the form of periodic structures (see Figure 1.10). They may exhibit a refractive index between zero and one (ultrarefraction) or negative (inversion of Descartes' law, see Figure 1.9). They are also known as *left-handed materials*.

Miniaturization of antennas stems from the action of reducing their volume, which has the consequence of rapidly reducing their effectiveness as well. Wheeler's work to qualify an antenna as *electrically short*, uses the criterion that is largest dimension d must be such that:

$$\frac{d}{\lambda} \leq \frac{1}{2.\pi} \quad [3.23]$$

where d, according to the work of Chu, is the radius of the so-called sphere of *less clutter*: it is the smallest sphere which encapsulates the antenna, excluding its power supply circuit. Chu, who extends Wheeler's relation between the volume of the sphere and the quality factor Q, divides the radiated field into spherical modes, and uses the sphere of less clutter to determine the minimum number of modes compatible with a given radiation.

For a linear antenna such as dipoles, miniaturization consists of decreasing its length, using a dielectric with high permittivity or folding it a certain number of times. In numerous cases – for instance, in VHF or LF applications – the antennas are used far beyond their resonance frequency. It is a question of the link budget and the sensitivity of the receivers. For planar antennas, such as microstrip antennas, we need to decrease their surface area.

Let us recap the formulae for the EFIE and MFIE, which express the electrical field and the magnetic field in the Fraunhofer zone:

$$\vec{E}(\vec{r}) = \frac{j.k^2}{4.\pi.\omega.\varepsilon}.\iiint_V \left[(\vec{J}(\vec{r}') \wedge \hat{u}) \wedge \hat{u} \right].\frac{e^{-j.k.r}}{r} dv$$

$$\vec{H}(\vec{r}) = \frac{j.k}{4.\pi} \iiint_V \left(\vec{J}(\vec{r}') \wedge \hat{u} \right).\frac{e^{-j.k.r}}{r} dv$$

They express the fact that the fields $\vec{E}$ and $\vec{H}$ are vectors resulting from the distribution of other vectors, which are the currents within a finite volume V. This results in a radiation pattern where the *purity* of the field components, in terms of the nominal component (copolarization) and cross component (crosspolarization), depends on the directional arrangement of the currents $\vec{J}$ within the volume V. The surface area available for the currents to circulate determines the directivity of the antenna (equivalent aperture). In numerous techniques, reducing the size of the antenna, and therefore the conductive surface area, consists of altering the way in which the currents circulate, forcing them to take a longer path whilst still reducing the size. Figure 3.27 illustrates a technique to lengthen the route taken by the surface currents by adding slots, which introduce new resonances at frequencies lower than the fundamental mode frequency. We need to prevent these slots from radiating – a task which, depending on how thin they are, may pose problems in terms of practical sensitivity during manufacture.

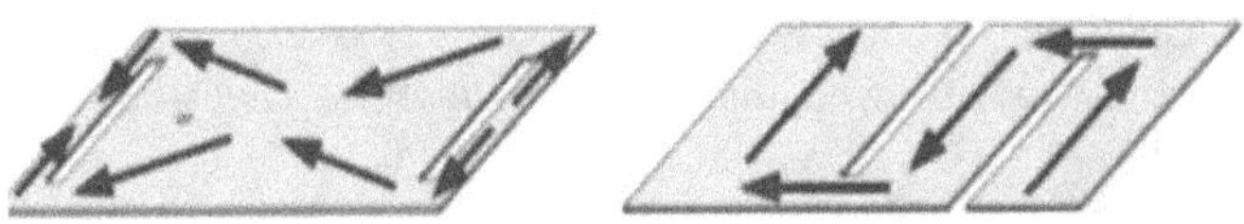

a) b)

Figure 3.27. *Decreasing antenna size by elongating the current lines by means of slots*

For a printed antenna, it is possible to short-circuit the patch at its center with a vertical metal wall (Figure 3.23), which connects it to the ground plane (a PIFA: *Planar Inverted-F Antenna*). This results in an antenna whose physical length is $\frac{\lambda_0}{4}$, representing a 50% reduction on its initial size. Techniques, such as with a PIFA, consist of engineering an overall increase in the antenna's effective inductance by introducing localized elements which affect the antenna's impedance, elongating the routes taken by the currents and thus decreasing the resonance frequency, equivalent to a reduction of the antenna's resonant dimension. It is also possible to take action on the dielectrics, which decreases the wavelength $\lambda_g = \frac{\lambda_0}{\sqrt{\varepsilon_r.\mu_r}}$ and can introduce a stark contrast in impedance between the dielectric material and the surrounding free space, given the storage of a large amount of electromagnetic energy in the so-called *near field* region.

The *fractal* technique is a miniaturization method (Falber 2000). The special feature of this method is that, by constant fracturations, with elementary and identical shape, it can squeeze the maximum possible amount of material into the minimum possible volume. In its shape, the antenna resembles the elementary pattern used for fractalization. The space gained in terms of volume means that other shapes can be put in place to obtain, for example, multi-band antennas. In addition, repetition of the patterns is easy to code.

In fact, the performance of an antenna is primarily tied to the materials used, which must exhibit minimal losses. Current techniques for miniaturization of antennas, as briefly indicated above, generally consist of charging the radiating component (dielectric materials, localized charges, introduction of slots or short-circuits) and taking action on the geometric of the structure (folding, specific shapes). The most compact structures often stem from a combination of the different techniques. In addition, miniaturization of an antenna comes with reductions in performance (reduced functional band, efficiency, loss of purity of polarization, increased sensitivity to the surroundings), as a result of the laws of physics.

The value $\frac{\lambda}{2.\pi}$ introduced as being the radius of the radiansphere by Wheeler corresponds to the transition distance between the near field, where the electromagnetic energy is stored in and around the antenna, and the far field, where it is propagated (section 2.12). Thus, this work, which is limited to the canonical cases of electrically short resonating antennas, can be used to establish the relation between the volume of the sphere occupied by the antenna and the product of its efficiency (η) by its bandwidth $\frac{\Delta f}{f_0}$:

$$\frac{\Delta f}{f_0}.\eta = \left(\frac{2.\pi.d}{\lambda}\right)^3 \qquad [3.24]$$

where $d << \frac{\lambda}{2.\pi}$.

The product of the bandwidth by the efficiency is thus directly linked to the equivalent volume of the antenna (the electrical length to the power of 3).

Pozar re-examined the fundamental limits of the minimum quality factor of an arbitrary, electrically short antenna. In particular, he focused on the impact of the polarization and the presence of TM, TE or TEM modes on the values of Q.

In summary, the enthusiasm sparked by the race for antenna integration, sadly, has theoretical limits. A reduction in size invariably results in a deterioration of the radioelectrical performances, particularly in terms of the yield and bandwidth. The quality factor Q associated with these elements has a minimum limit which becomes higher as the antenna's size (in λ) becomes smaller. The bandwidth Δf, given by equation [3.12], restated below, for a given VSWR, shows the increasingly selective nature of miniature devices:

$$\frac{\Delta f}{f_0} = \frac{VSWR - 1}{Q.\sqrt{VSWR}}$$

In addition, a high quality factor leads to a high concentration of the EM field in the vicinity of the antenna, causing high sensitivity to losses through the Joule effect and in the dielectrics, which may cause dramatic drops in yield. Also, extending the bandwidth is one of the problems linked to compact antennas which has attracted most research effort. What techniques should we employ to make the best possible use of the space around the antenna, and increase its effective volume? It is always possible to lower the resonance frequency by adding an external capacitor or inductor, but this does not alter the quality factor of the resonance, and therefore the bandwidth. At an equal resonance frequency, the shortest antenna will have the lowest bandwidth. This may pose a problem when the antenna is intended to be used for an application for which a frequency range spanning tens of MHz has been allocated, or for multi-band applications. The existing literature (Skrievervik 2011) provides formulae giving theoretical limits in terms of minimum quality factor Q_{min} and maximum gain G_{max} for a short antenna with rectilinear polarization:

$$Q_{min} = \frac{1}{(k.d)^3} + \frac{1}{k.d} \tag{3.25}$$

$$G_{max} = (k.d)^2 + 2.k.d \tag{3.26}$$

4

Characteristic Parameters of an Antenna

Jean-Pierre BLOT
EUROSATCOM, Les Ulis, France

4.1. Characteristic parameters of an antenna[1]

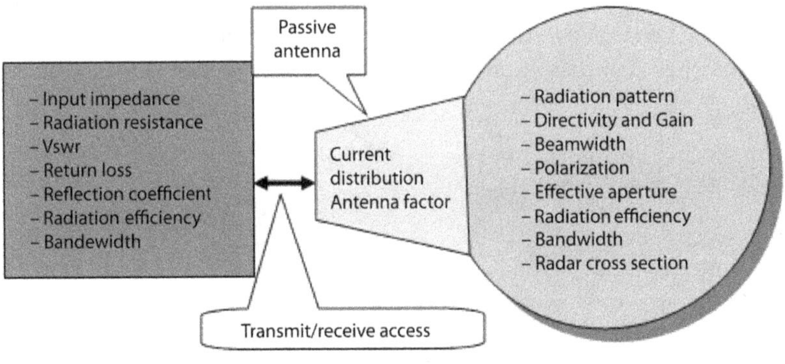

Figure 4.1. *Main parameters of an antenna*

The main parameters which characterize an antenna are shown in Figure 4.1. In the center are the accesses, defined by Figures 3.14a and 3.14b; to the right are the radioelectrical parameters; to the left, the electrical parameters. For those parameters

---

For a color version of all figures in this book, see www.iste.co.uk/favennec/electromagnetic2.zip.

1 (Balanis 2008; Huang 2008; Milligan 2005).

Electromagnetic Waves 2,
coordinated by Pierre-Noël FAVENNEC. © ISTE Ltd 2020

not discussed here, readers may refer to the pre-existing literature – in particular, to Balanis (2008).

4.1.1. *Capture surfaces or equivalent surfaces on an antenna*

Certain parameters need defining precisely in order for readers to properly understand how they are used. Such is the case of the capture surface, which does not only apply to aperture antennas.

Consider a receiver antenna, located in the far field of a transmitter. If the dimensions of the receiver antenna are small in comparison to its distance from the transmitter, the receiver antenna can be deemed to be stimulated by a plane wave with a constant power density P (W/m^2).

From the standpoint of circuit theory, the receiver antenna can be viewed as a generator depending on a voltage or current, depending in turn on whether we use the Thévenin equivalent or Norton equivalent (sections 1.3.1.1 and 1.3.1.2). The reciprocity theorem tells us that the impedance of the antenna is the same in both transmission and receiving modes, and that the maximum power available for reception and that transmitted is obtained when the impedance of the transmitter or receiver is equal to the complex conjugate of the antenna's internal impedance Z_A*. We can then define the equivalent capture surface A_e (effective aperture) of an antenna as being the ratio of the maximum power available for reception to the incident power density P:

$$A_e(\theta,\varphi) = \frac{P_{Av}(\theta,\varphi)}{P}\,(\text{m}^2) \qquad [4.1]$$

This capture surface depends, of course, on the antenna's orientation in relation to the incident wave. In practice, we often look at its maximum value, when the antenna's orientation is such that it captures the maximum power in a particular direction (θ_0,φ_0). This is the value $A_e(\theta_0,\varphi_0) = \frac{P_{Av}(\theta_0,\varphi_0)}{P} = A_{e.MAX}$.

As outlined in sections 1.3.1.1 and 1.3.1.2, in the case of maximum power transfer:

$$A_{e.MAX} = \frac{|V_T|^2}{8.P(R_L + R_r)}.\eta_r \qquad [4.2]$$

where $\eta_r = \dfrac{R_r}{R_L + R_r}$ is the radiation loss factor, similarly with the losses in the different resistors R_T and R_L.

The total equivalent surface, taking account of other losses in the resistors (Figures 1.11 and 1.12), is:

$$A_{e.MAX} = \frac{|V_T|^2}{8.P(R_L + R_r)} \cdot \left[\frac{R_T + R_r + R_L}{(R_L + R_r)^2}\right] \quad [4.3]$$

which is an important parameter in characterizing an antenna. This maximum equivalent surface enables us to define the directivity of an antenna by the general relation $A_e = \dfrac{\lambda^2}{4.\pi}.D_0$. There is always a particular direction in which the radiation is maximal. Note here that A_e is not based on any of the antenna's geometric characteristics. It is only with parabolas, horns and hollow waveguides that the capture surface is directly linked to the geometric surface area intercepted by the incident plane wave. Thus, in other cases, such as wire antennas, A_e cannot be tied to any particular geometric surface.

For example, the Hertz dipole (Balanis 2008) has an equivalent surface $A_e(\theta,\varphi) = \dfrac{3.\lambda^2.\sin^2(\theta)}{8.\pi}$.

It is independent of the dimensions of the dipole, but a function of the frequency. Its maximum value is:

$$A_{e.MAX} = \frac{3.\lambda^2}{8.\pi} \quad [4.4]$$

4.1.2. *Directivity and gain*

Consider a link between a transmitter E and a receiver R placed in a vacuum in one another's far field, at a distance d apart, such that the polarizations of the waves transmitted and received are matched, meaning that they are aligned. If the transmitter E were equipped with an isotropic antenna (which does not have a physical existence), the power flux P_E exiting a sphere, whose center is E and whose radius is d, would be uniformly distributed across that sphere. Per unit surface, the power flux would be:

$$\frac{P_E}{4.\pi.d^2} \qquad [4.5]$$

As the antenna of the transmitter, in the direction of the receiver, presents an absolute gain G_E, the power flux density in that direction is:

$$\frac{G_E.P_E}{4.\pi.d^2} \qquad [4.6]$$

The product $P_E.G_E$ is the apparent radiated power (ARP) or the equivalent isotropically radiated power (EIRP). The receiver antenna with effective surface A_e takes the following power from the wave:

$$A_e.\frac{G_E.P_E}{4.\pi.d^2} \qquad [4.7]$$

The power received at the input to the receiver is therefore:

$$P_r = P_E.\frac{G_E.S_R}{4.\pi.d^2} \qquad [4.8]$$

To make this easier to understand, we set $S_R = A_e$ for the effective surface of the receiver.

The absolute gain of the antenna (gain in comparison to an isotropic antenna) and its effective surface are linked by the relation:

$$G = \frac{4.\pi}{\lambda^2}.S$$

N.B. The gain G is equal to the directivity D minus the losses (in dB). Here, $S = A$, the effective surface takes account of the losses, so it is smaller than the physical surface of the antenna. The expression [4.8] can take two different forms, depending on whether we are expressing it as a function of the gain or of the antennas' effective surface:

$$\frac{P_R}{P_E} = G_E.G_R.\frac{\lambda^2}{(4.\pi)^2 d^2}$$
$$\frac{P_R}{P_E} = S_E.S_R.\frac{1}{\lambda^2.d^2} \qquad [4.9]$$

The power attenuation:

$$a_{dB} = 10.\log_{10}\left(\frac{P_E}{P_R}\right) = 22 + 20.\log_{10}\left(\frac{d}{\lambda}\right) - G_E - G_R$$

The first expression of formulae [4.9] is the Friis expression in its simplest form (the ideal case, no multiple paths, no attenuation other than that in free space).

4.1.3. *Relation between gain, directivity and radiation pattern*

The gain of an antenna in a direction (θ,φ) is the ratio between the power radiated per unit solid angle $P(\theta,\varphi)$ in that direction and the power radiated by an isotropic source, fed by the same total power. Such a source has no physical meaning because, physically, there is at least one zero in the radiation pattern, where the polarization closes the sphere encapsulating the antenna.

If P_E is that total power, the power radiated by the isotropic source in the unit of solid angle is:

$$\frac{P_E}{4.\pi}$$

The directivity of an antenna in a direction (θ,φ) is therefore defined by:

$$D_R(\theta,\varphi) = \frac{P(\theta,\varphi)}{\dfrac{P_E}{4.\pi}} \qquad [4.10]$$

In the specific case where the antenna radiates completely the power supplied to its access points in transmission mode, or recovers that energy when operating in receiver mode (i.e. the lossless case), the total power P_T is given by:

$$P_T = \int_0^{4\pi} P(\theta,\phi) d\Omega \qquad [4.11]$$

where $d\Omega = \sin(\theta) d\varphi . d\theta$.

The directivity now depends only on the variation of $P(\theta,\varphi)$, and therefore on the radiation pattern, because $P(\theta,\varphi) = P_{MAX} . r(\theta,\varphi)$, so in the direction of maximum radiation (θ_0, φ_0):

$$D_0 = \frac{4.\pi}{\int_0^{4.\pi} r(\theta,\varphi) d\Omega} \qquad [4.12]$$

In decibels, $D_0(dBi) = 10.\log_{10}(D_0)$ are indeed dB with respect to the isotrope (equation [4.6]). The gain G_0 is equal to the directivity less the internal losses in the antenna, so:

$$G_0 = D_0 . \eta \qquad [4.13]$$

$0 \leq \eta \leq 1$ is the yield of the antenna.

We also have the radiation pattern given by the ratio of the equivalent surface as a function of the observation point $S(\theta,\varphi)$ to its maximum $S_{MAX} = S(\theta_0,\varphi_0)$:

$$r(\theta,\varphi) = \frac{S(\theta,\varphi)}{S_{MAX}} \qquad [4.14]$$

4.1.4. *Effective height or effective length*[2]

In the case of wire antennas, the effective height or effective length h_{eff} (in meters) can be used in place of the capture surface (effective surface):

2 (Rudge 1982).

$$\left|\vec{E}_{inc}\right|.h_{eff} = V_{co} \qquad [4.15]$$

$$I_{in}.h_{eff} = \int_0^h I(z).dz \qquad [4.16]$$

where V_{co} is the voltage induced at the terminals of the antenna in an open circuit by the electric field $\vec{E}_{inc}$ in the vicinity of the antenna, I_{in} is the current at the base of the antenna and $I(z)$ is the current distribution along the antenna. Figure 4.2 illustrates the principle of effective height; the surfaces under the distribution curves of current are equal.

The effective height is defined as the ratio of the antenna's open-circuit voltage V to the incident field striking it:

$$h_{eff} = \frac{V}{E} \propto \sqrt{A_{eff}} \, ...(m) \qquad [4.17]$$

In summary, an antenna's effective height is the length of a fictitious wire antenna which is traversed by a constant current, equal to the average current in the real antenna, and radiating the same level of power.

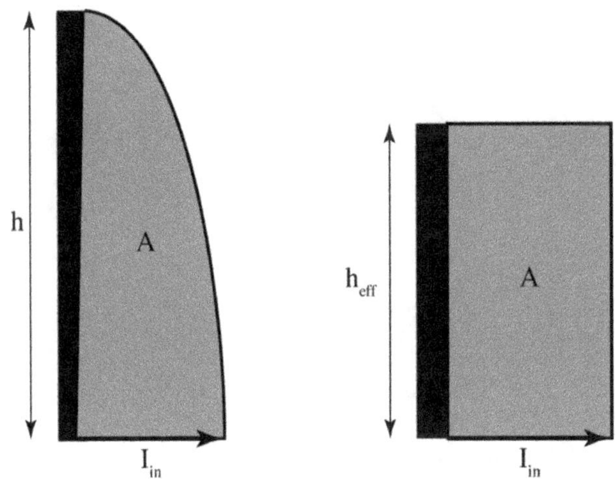

Figure 4.2. *Current distribution I(0) = I_{in} and effective height h_{eff} of a quarter-wave antenna*

4.2. Link budget

A transmitter is characterized by its power, and a receiver by its sensibility. To ensure proper reception, the carrier to-noise ratio (C/N) of the signal, measured at a certain point on the reception stage after demodulation, must be greater than a given threshold. This ratio is frequently expressed as a ratio of energy $\frac{E_b}{N_0}$ where E_b is the energy of a transmitted chip, and where N_0 (in W/Hz) denotes the spectral noise density. In the absence of any signal, the thermal agitation of the electrons in a conductor produces white noise (the Johnson effect), with a spectral power density $N_0 = k.T$ in the radioelectric domain, i.e. at high frequencies, where k is the Boltzmann constant ($k = 1.38.10^{-23}$ joules/Kelvin) and T is the noise temperature (in Kelvin). For a modulation of the spectral breadth Δf (the Nyquist band), the noise power is approximately equal to $w = N_0.\Delta f$. Note that the bandwidth Δf of a receiver is its equivalent noise band. As a minimum, it is the bandwidth required by the modulation used. In a real scenario, the effective noise temperature at reception must take account of all forms of noise between the transmitter and the receiver (natural noise, industrial noise, etc.).

The relation between the $\frac{C}{N}$, i.e. the level of the carrier with the total noise power N, and $\frac{E_b}{N_0}$ is given by:

$$\frac{C}{N} = \frac{E_b.D_u}{N_0.\Delta f} \quad [4.18]$$

In decibels:

$$\left(\frac{C}{N}\right)_{dB} = \left(\frac{E_b}{N_0}\right)_{dB} + 10.\log_{10}\left(\frac{D_u}{\Delta f}\right)$$

where:

- C is the power of the carrier received (in watts);
- N is the noise power in the band Δf of the channel (in watts);
- Δf is the width of the channel (in hertz);
- D_u is the usable datarate transmitted per second (in Baud).

The usable datarate transmitted D_u in a channel whose width is Δf is given by $D_u = r.D$, with D being the transmittable datarate in a channel whose width is Δf and r being the yield of the code.

For a perfect demodulator, each type of modulation is characterized by a theoretical Bit Error Rate (BER) as a function of the operational threshold $\frac{E_b}{N_0}$. For example, for a BPSK modulation for a BER of 10^{-8}, the operational threshold is $\frac{E_b}{N_0} \approx 12$.

Equation [4.18], by the value of *C*, gives the power of the transmitter via the propagation channel based on the Friis formula (equation [4.9]) or by equation [4.26] below.

The *sensitivity S* is the minimum level of power *C* for which the ratio $\frac{E_b}{N_0}$ is greater than the operational threshold of the receiver. Thus, on a logarithmic scale, we have:

$$S_{dB} = \left(\frac{E_b}{N_0}\right)_{LIMIT.dB} + N_{dB} \qquad [4.19]$$

with:

$$N_{dB} = 10.\log_{10}\left(kT_{system}.\Delta f\right) + N_{receiver}$$

where:

$$N_{receiver} = 10.\log_{10}\left(k.T_0.\left(F_{receiver} - 1\right).\Delta f\right)$$

where $N_{receiver}$ is the noise power expressed at the input of the receiver due to its noise factor $F_{receiver}$ and where $T_0 = 290$ K, the reference temperature (Friis 1944).

The sensitivity of a receiver, therefore, is equal to the minimum power necessary to ensure a certain quality of service, or a certain datarate, in the presence of the noise alone. A receiver itself must not produce more “noise” than the antenna received.

Thus, the sensitivity of a receiver defines its capacity to receive weak signals. It is expressed in a variety of ways, depending on the application at hand:

– according to ITU-R SM. 1840, it is the minimum level (in μV, dBμV or dBm) that must be achieved by the signal applied at the input to the receiver, in order to be demodulated and heard as output;

– in data transmission and digital radio, it is the level of antenna input giving a BER lower than a given value;

– for a receiver entering into a system (example: a wireless beam, a satellite repeater, etc.), the sensitivity is divided into two parameters: the input noise factor and the overall gain.

4.3. Power and noise temperature[3]

In a link budget, we use the formula to find the noise power N available at the *reception* access of an antenna. Indeed, it is when the antenna is in receiver mode that the noise generated by different sources becomes important in determining the quality factor (G/T), particularly in applications relating to *satellites*. Any signal source generates a certain amount of noise at the same time as the signal. The receiving equipment working on that signal adds a further measure of noise, known as *internal noise*, which must be differentiated from the *external noise* picked up by the antenna.

For internal noise, we have three main categories:

– thermal noise;

– shot noise;

– scintillation noise.

Only thermal noise, known as the Johnson effect, is taken into account, as expressed above, with $w = k.T..\Delta f$. This noise power is due to the antenna's ohmic losses, up to the input of the receiving amplifier, including the amplifier's own noise power, expressed at its input. The reception amplifier, by its gain, must mask the noise of microwave and electronic components downstream, hence the advantage of putting it as close as possible to the antenna located upstream.

3 (Thourel 1956).

Noises external to the antenna refer to exterior noise picked up by the antenna. Note that, to begin with, if that antenna were placed in a perfectly absorbing chamber (a black body), and taken to the absolute temperature T_a, the noise power received by the antenna would be equal to:

$$W = k.T_a.\Delta f \tag{4.20}$$

but the space surrounding the antenna is not at a uniform temperature. The radiation pattern will never be isotropic. W will therefore depend on the regions which are "perceived" by the radiation pattern in the receiver's bandwidth Δf.

T_a, known as the *noise temperature of the antenna*, does not depend on Δf, so we can characterize the latter. However, T_a depends on the radiation pattern. Thus, the noise temperature depends on the direction in which the antenna is turned. Hence, T_a is not an intrinsic characteristic of the antenna.

In the case of an antenna pointed at high sites, if the level of the secondary lobes directed at the ground is very low, little noise power will be picked up, and T_a will be low: around 20 K. We say that the antenna is *cold*. The antennas of ground base stations are among so-called cold antennas. We have two main types of antennas, one colder than the other, depending on the positioning of the primary source, as shown in Figure 4.3.

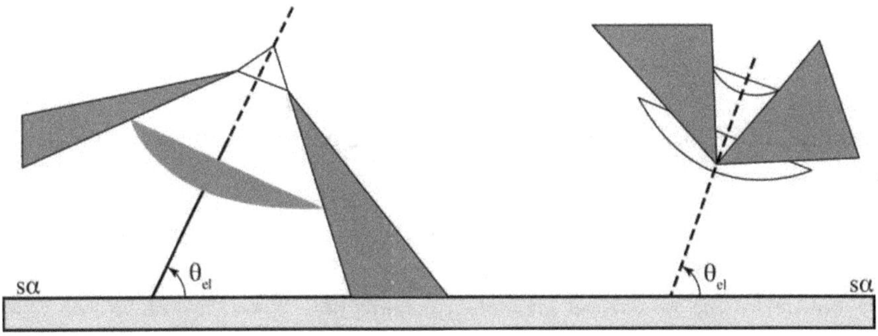

Figure 4.3. *Cold antennas*

Indeed, a horn exhibits a very broad radiation pattern, and a portion of that radiation spills over the primary or auxiliary reflector. Depending on the direction of the horn, that spill-over will, in one case, be channeled into the ground at a temperature of around 290 K, and in the other, directed into the sky at a temperature varying by $T_{sky} = T(90°)/\sin(\theta_{el})$ (ratio reference CCIR 720-1 and curves in

Figure 1.1), where $T(90°)$ is the temperature in K, for a clear sky at elevation $\theta_{el} = 90°$. For example, at $\theta_{el} = 30°$, for $f = 10$ GHz, $T_{sky} = 15$ K.

Note, briefly, that in a similar way to internal noise, the external noise picked up by the antenna is given by equation [4.20]. To demonstrate this, we need to delve into the realm of thermodynamics. At a temperature T, all bodies emit or absorb thermodynamic radiation, the power of which is a function of T and of the wavelength λ. A body which absorbs all of the incident power is known as a *black body*. Yet even a black body, in turn, radiates some of the energy. It would be invisible if it absorbed all light energy. Thus, it absorbs a certain amount of power P_a, which is a portion of the incident power P_i.

The ratio $A = \frac{P_a}{P_i}$ is known as the body's *absorbing power*. For example, in the case of attenuation α due to ohmic losses, $A = (1 - \frac{1}{\alpha})$. When a body emits energy, we say that it has *emissive power*, also known as *brightness*, which is the power radiated per unit solid angle and per unit surface of the body, in the direction normal to that surface. If B is the emissive power, B is expressed in watts per unit solid angle and per unit surface area of the body. By virtue of reciprocity, if W is the power radiated by the black body at temperature T, the emissive power of a body at the same temperature with an absorbing power A is:

$$B = A.W \tag{4.21}$$

In other words, if a medium absorbs A% of the thermodynamic energy provided by a noise source, it will reciprocally emit A% of the energy that a black body, at the same temperature, would radiate.

If we write $B = A.W$, this indicates that a passive body emitting power can only be absorbent. Indeed, if $A = 0$, then we have $B = 0$. Hence, a metal reflector that is perfectly reflective (because it is perfectly conductive: $A = 0$) can radiate no power of its own. In addition, a waveguide placed between a noise source and a receiver does not increase the noise temperature unless that guide has losses α, in which case the noise temperature is increased by:

$$T_{waveguide} = T_{Physical} \cdot \left(1 - \frac{1}{\alpha}\right)$$

For example, an attenuation of 0.1 dB produces a temperature increase of 7 Kelvin for an ambient temperature of:

$$T_{Physical} = 273\ K + 27°C = 300\ K$$

Thus, all external attenuators such as the atmosphere (rain, clouds, etc.) produce a noise power given by equation [4.21], including the ohmic losses due to the finite conductivity σ or dielectrics of the reflectors and the constitutive elements of the primary source of the antenna.

4.3.1. *Noise temperature received by an antenna*[4]

The reference work demonstrates that the external noise temperature expressed by the radiation pattern of the antenna is given by the known formula:

$$T_a = \frac{1}{4.\pi} \int_{\Omega} G\theta,\varphi).T(\theta,\varphi).d\Omega \qquad [4.22]$$

In practice, we look at the radiation pattern $r(\theta,\varphi)$. We have seen that the gain $G(\theta,\varphi)$ is given by the formulae [4.12] and [4.13]. Hence, T_a will always be calculated by the formula:

$$T_a = \frac{\int_{\Omega} r(\theta,\varphi).T(\theta,\varphi)\, d\Omega}{\int_{\Omega=4\pi} r(\theta,\varphi)\, d\Omega} \qquad [4.23]$$

Consider the formula $W = k.T_a.\Delta f$ which, at the access to the receiver, gives the total noise power picked up by the antenna:

$$W = k.T_{System}.\Delta f \qquad [4.24]$$

where T_{System} is the sum of the temperature caught by the antenna by its radiation pattern and the internal temperature due to the different ohmic losses of the antenna's hyperfrequency components.

4 (Thourel 1956).

4.3.2. *Link budget and Friis formula*

The Friis equation [4.9] is also known as the telecommunications equation in its simplest form, with the ideal case where the antennas are visible to one another, with no multiple paths or other disturbances of the ambient environment, be they atmospheric, smoke, dust, etc. The equation is recapped below:

$$\frac{P_R}{P_E} = G_E.G_R.\left(\frac{\lambda}{4.\pi.d}\right)^2 \quad [4.25]$$

where :

– P_E is the power in watts (W) transmitted by a satellite, for example, supplied to the receiver antenna;

– P_R is the power in watts (W) received by the receiver antenna;

– G_R is the gain of the receiver antenna *in the direction of the transmitter antenna if it is visible*;

– G_E is the gain of the transmitter antenna *in the direction of the receiver antenna if it is visible*;

– λ is the wavelength at the working frequency.

With equation [4.25], we assume that the antennas are visible to one another, and are correctly aligned in terms of direction and polarization of the EM field. In reality, these conditions are never met because of obstacles, reflections, multiple journeys, rain, etc. The most realistic equation must take account of all these factors degrading the link, including improper alignment which loses its meaning if the antennas are not visible to each other:

$$\frac{P_R}{P_E} = G_E.G_R.\eta_E.\eta_R.L_{diverse}.\left(1-|s_{11}|^2\right).\left(1-|s_{22}|^2\right).\left|\vec{u}.\vec{v}^*\right|^2.\left(\frac{\lambda}{4.\pi.d}\right)^2 \quad [4.26]$$

where:

– η_E is the performance coefficient of the transmitter antenna;

– η_R is the performance coefficient of the receiver antenna;

– s_{11} is the reflection coefficient on the transmitter antenna;

– s_{22} is the reflection coefficient on the receiver antenna;

– $\vec{u}$ is the linear or circular polarization of the wave received from the transmitter;

– $\vec{v}^*$ represents the conjugation of the linear or circular polarization of the receiver antenna;

– $L_{diverse}$ represents the diverse losses in the propagation channel due to the atmospheric agents for example, or to multiple journeys.

In order to obtain as reliable a prediction as possible, it is important to define the reference access of the antennas: transmission access and receiving access in a set of technical specifications.

In order for the Friis equation [4.25] to be applicable, it is necessary not only for there to be no obstacle on the straight segment whose length d is linking the transmitter to the receiver, but there must also be a free environment. More precisely, the first Fresnel ellipsoid, the focal points of which are the antennas and the smaller diameter of which is $0.5\sqrt{\lambda.d}$, must contain no obstacles, which is almost never the situation in the radiomobile context of macro-cellular technology, where the signal is affected by numerous possible distortions, from the Doppler effect to Rayleigh fading of amplitude and phase on the multiple paths, lack of direct visibility between the antennas, etc. Multiple methods and models of varying degrees of complexity (and therefore varying degrees of accuracy) have been developed to determine the loss of propagation in complex environments. These methods are discussed in Chapter 2 of Volume 1.

Equation [4.26], however, shows the importance of the electrical and radioelectrical characteristics of the antennas, particularly in the realm of spatial activities, to control the losses in the constitutive circuits, including both receivers and transmitters.

4.4. Quality factor Q = G/T

We shall limit our discussion here to the case of a single-carrier satellite downlink because of the availability of a simple equation for the link budget, given by the Friis equation.

The quality factor of a link is defined by the antenna's receiver gain G_R and its system temperature T_{System} at the receiver access point. Consider:

$$Q = 10Log_{10}\left(\frac{G_R}{T_{System}}\right)(\text{DB/ K}) \tag{4.27}$$

As stated previously, T_{System} is the noise temperature of the antenna system – i.e. the sum of the external noises captured by the radiation pattern, the internal noises generated by all the ohmic losses of the constitutive elements of the antenna and the noise created by the receiver low-noise amplifier (LNA). For a satellite link (CCIR reference 552-3):

$$T_{system} = T_a + T_{receiver}$$

so:

$$T_{system} = \frac{L_i - 1}{\alpha L_i}\left(T_{atmos} - T_{sky}\right) + \frac{1}{\alpha}\left[T_{sky} + T_{ground} + (\alpha - 1).T_{phys}\right] + T_{receiver} \tag{4.28}$$

with:

– T_{sky}, the temperature of the atmosphere with a clear sky, calculated by $T_{sky} = T(90°)/\sin(\theta_{el})$ (reference of CCIR report 720-1);

– T_{ground}, the temperature contributed by the ground (through rain if applicable), measured or calculated by equation [1.39];

– T_{atmos}, the physical temperature of the atmosphere and precipitation;

– T_{phys}, the physical temperature of the non-radiating subsets between the antenna and the LNA (waveguides, coaxial cables, orthomodes, etc.) of the primary source;

– $T_{receiver}$, the noise temperature of the receiving equipment;

– $\alpha \geq 1$, the ohmic losses due to the non-radiating elements, radomes, etc.;

– L_i, the additional attenuation due to the atmosphere in particular conditions (regions with a temperate or tropical climate) and to precipitation.

It is helpful to write equation [4.28] in the following form:

$$T = T_a^{'} + \Delta T_a^{'} + T_r \quad [4.29]$$

where $T_a^{'}$ is the antenna temperature in clear weather ($L_i = 0$ dB):

$$T_a^{'} = \frac{T_{sky} + T_{ground}}{\alpha} + \frac{\alpha - 1}{\alpha} T_{phys} \quad [4.30]$$

$\Delta T_a^{'}$ is the additional noise temperature of the antenna due to atmospheric attenuation (clouds, etc.) and precipitation:

$$\Delta T_a^{'} = \frac{L_i - 1}{\alpha L_i}\left(T_{atmos} - T_{sky}\right) \quad [4.31]$$

We assume that the gain of the LNA which is upstream of the telecom equipment (see Figure 4.4) is high enough to mask their noise power brought back to the entrance of the LNA because, by means of a simple calculation:

$$T_R = T_{LNA} + \frac{T_2}{G_{LNA}}$$

where T_2 is the noise temperature of the equipment downstream of the LNA.

Why is this factor so important? To simplify the explanation, we take the Friis equation which is valid in the case of antennas that are visible to one another (space applications, radio-relay systems, etc.). The explanation then becomes immediately apparent when we mathematically express what was said to begin with. Indeed, we have seen that the power of the signal received at the receiver antenna access is given, based on formula [4.9], by:

$$P_R = P_E.G_E.G_R \frac{\lambda^2}{(4\pi)^2 d^2} \quad [4.32]$$

where $P_E.G_E$ is the EIRP (equivalent isotropically radiated power), $\frac{\lambda^2}{(4\pi)^2 d^2}$ the loss in free space due to the spherical nature of the wavefront of the EM incident wave coming from the transmitter – here, a satellite.

The received noise power is given by $N = k.T_{System}.\Delta f$. By finding the signal-to-noise ratio (C/N), setting $P_R = C$ for notation consistency:

$$\frac{C}{N} = \frac{[EIRP]_{satellite}}{k.\Delta f}\left[\frac{\lambda}{4.\pi.d}\right]^2 \cdot \left[\frac{G_R}{T_{system}}\right] \quad [4.33]$$

Equation [4.33] is written:

$$\frac{C}{N} = \gamma.\left[\frac{G_R}{T_{system}}\right] \quad [4.34]$$

The factor γ is fixed by the characteristics of the satellite and the propagation conditions. If we want to achieve a specific C/N, our only option is to adjust the quality factor $Q = \frac{G_R}{T_S}$ which characterizes the ground station.

For terrestrial links, through the model of the propagation channel which is immutable, it is possible to act, for a given environment, on the transmitter's electrical and radioelectric characteristics, on its height, etc., on its EIRP variable, whereas the receiver (the handset) has a sensitivity fixed by design.

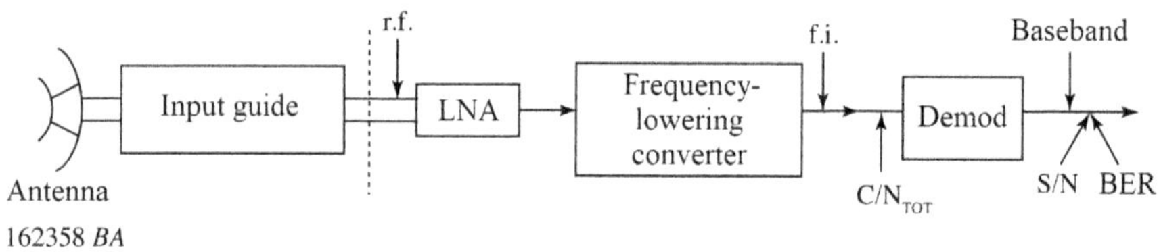

Figure 4.4. *Diagram showing the principle of a receiver chain*

5

Digital Methods

Jean-Pierre BLOT
EUROSATCOM, Les Ulis, France

5.1. Introduction to digital methods[1]

5.1.1. *Overview of the main digital methods*

Numerous published articles and books focus on CEM methods. In addition to the works listed in the bibliography, readers could usefully consult the article by P. Sumithra and D. Thiripurasundari, "A review on Computational Electromagnetics Methods", in the journal *Advanced Electromagnetics* (Sumithra and Thiripurasundari 2017), and the article "Computational electromagnetics" on Wikipedia.

Electromagnetic simulators are mainly used by antenna engineers, and more generally, those working in the microwave domain. Work begins by analyzing the physical structure (i.e. the geometry and electrical properties of the materials involved), and continues with a complete overview of the properties of the devices and circuits (including propagation, radiation, parasitic effects, etc.). For most problems, no solution can be found analytically, and a digital approach is necessary. However, it is no easy task to write a computer code to solve a class of problems. Similarly, real expertise in the field is needed in order to correctly interpret the results returned by these codes. Such software, which is generally highly sophisticated, requires computers with extremely powerful central processing units (CPUs) to produce accurate results in a reasonable timeframe. An effective and

For a color version of all figures in this book, see www.iste.co.uk/favennec/electromagnetic2.zip.

1 (Yee 1966; Harrington 1993; Tapan 2000; Balanis 2007; Picon 2009).

Electromagnetic Waves 2,
coordinated by Pierre-Noël FAVENNEC. © ISTE Ltd 2020

precise computer simulation of the different problems encountered in CEM is made possible by carefully designed digital techniques. The codes enable us to view the results on screen, in many cases providing more information than could be gleaned by laboratory tests or *in situ* tests, for a lower cost and with greater efficiency. In view of the constant increase in demand for high-performing electronic components, the industry is increasingly making use of digital simulations, before anything is actually built. It was in the 1960s that a new kind of analysis in the field of electromagnetics emerged, referred to as Computational Electromagnetics (CEM). The main pioneers in this domain were Kane S. Yee in the temporal domain and R.F. Harrington in the frequency domain. They demonstrated that numerical solutions to Maxwell's equations were capable of accurately describing the electromagnetic behavior of objects such as antennas without the need for prior experimentation. There has been a major shift towards digital simulation because of the exploitation of the microwave frequency band, with the mass development of wireless applications thanks to the miniaturization of electronic and hyperfrequency components and, in particular, of antennas.

In today's world, all CEM modeling software uses digital techniques to solve Maxwell's equations expressed in the form of integrals or differential equations. Figure 5.1 illustrates the most usual methods employed in the field of antennas.

The technique used and the nature of the equations needing to be solved have a crucial impact on the relevance of a particular software program to analyze a given problem, as do the computational resources available for the task: do we choose a frequency-domain analysis or a time-domain analysis? For the frequency domain, we obtain a response at a given frequency – i.e. at a specific time. For the time domain, we obtain the result over an extensive band of frequencies. Frequency-domain techniques are more effective in the modeling of narrow-band problems, or those with a high quality factor Q. Time-domain techniques are more advantageous for broadband structures whose dimensions are larger in relation to the wavelength.

CEM modeling techniques depend on the way in which Maxwell's equations are expressed:

– in the form of an integral which describes the behavior of the EM field and the currents on the enclosed surface of structures of any given shape and dimension;

– in the form of differential equations that describe the behavior of the EM field and currents at a point on the structure and in the space.

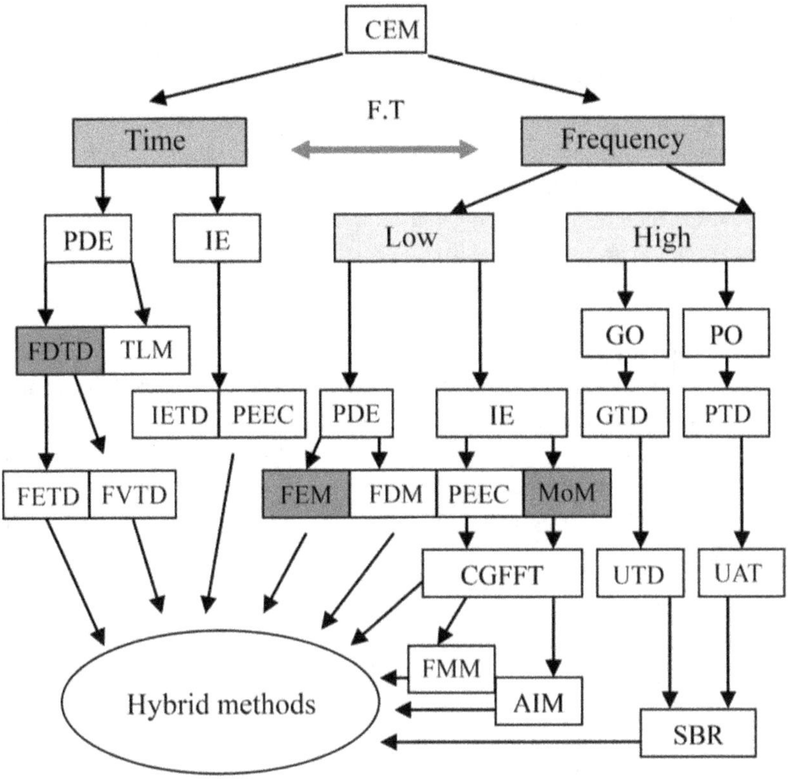

Figure 5.1. *Main digital methods used by electromagnetic simulators (extract from Er-Ping 2009)*

NOTES ON FIGURE 5.1.– *AIM*: Adaptive Integral Method; *CEM*: Computational Electromagnetic Methodologies; *CGFFT*: Conjugate Gradient Method with FFT; *FDM*: Finite Difference Methods; *FDTD*: Finite-Difference Time-Domain; *FEM*: Finite Element Method; *FETD*: Finite-Element Time-Domain; *FMM*: Fast Multipole Method; *FVTD*: Finite-Volume Time-Domain; *GO*: Geometric Optics; *GTD*: Geometric Theory of Diffraction; *IE*: Integral Equation; *IETD*: Integral Equation Time Domain; *MoM*: Method of Moments; *PDE*: Partial Differential Equations; *PEEC*: Partial Equivalent Element Circuit; *PO*: Physical Optics; *PTD*: Physical Theory of Diffraction; *SBR*: Shooting and Bouncing Ray; *TLM*: Transmission Line Matrix; *UAT*: Uniform Asymptotic Theory of Diffraction; *UTD*: Uniform Theory of Diffraction.

It is possible to switch from the expressions of Maxwell's equations in the form of integrals to those in the form of differential equations by reducing the radius of the closed curve delimiting a surface to near zero and the closed surface delimiting a volume to near zero. The two forms of expressions describe the same things, but produce fundamentally different CEM modeling techniques. Techniques based on the partial differential equation (PDE) form of Maxwell's equations analyze the geometry of a problem by sampling the entirety of the surface and volume of the structure. To calculate the EM far field, absorbent boundaries are used (PML for FDTD, for example). For the integral form, only the surfaces and boundaries between the regions are sampled, necessitating an often-laborious effort to find the Green's function which establishes the relation between an element of current J_S on the conductors and the EM field it creates. The Green's functions are determined by solving Maxwell's equations in each of the propagation media and solving the relations of field continuity at the interfaces. For example, for a 3D structure such as a printed patch antenna, the structure is sampled only on the conductive part of the patch. For this reason, the problem can be reduced to a 2D problem by finding the currents.

All digital techniques, regardless of their complexity, exhibit advantages but also drawbacks. Thus, it is still crucially important to conduct experiments and produce a model in order to validate the results yielded by the software. The algorithms used are typically classified as exact or low-frequency methods and approximate or high-frequency methods, and subclassified into *time*-domain and *frequency*-domain methods (see Figure 5.1). This classification depends on the ratio between the geometric dimensions of the problem and the wavelength at the frequencies in question. These subclassifications seem superfluous, because a solution across a broad range of frequencies can be obtained by a Fourier transform of a time solution. Similarly, a time solution can be obtained by the inverse transform of a frequency response. There are benefits and drawbacks to each of the domains. Note, in particular, that:

– the frequency domain has the advantage of being more efficient for frequency-by-frequency calculations over a narrow band, for dispersive media and media with loss, because the values of the parameters are fixed during the calculations. However, frequency methods are not effective for characterization across a wide range of frequencies. They are truly not suitable for nonlinear, non-stationary problems;

– the time domain has the advantage of being able to characterize a device across a wide range of frequencies, using the Fourier transform. It enables us to obtain a complete response for the transitory solution, which is particularly useful in the case of UWB (Ultra-Wide Band) devices. In the time domain, we can relatively easily introduce models of nonlinear components, non-stationary and heterogeneous components or materials. There is no matrix inversion but, briefly, time methods occupy a very important place in memory, in the case of a fine mesh to describe complex structures in detail, particularly in the presence of dispersive media and in the case of structures with a high quality factor Q.

All digital methods have one thing in common: they require the discretization of the variables and values in order to compute an estimation of them. It is the discretization of the problem which takes up computer memory, requiring high-performance clusters, vectorized and/or parallel processors, in order to (amongst other operations) invert large matrices to calculate the weights of the basic functions, such as in the method of moments (MoM). Indeed, the computer can only perform the operations step by step, but can perform an enormous number of such steps. These operations, finite in number, constitute the steps in the discretization. As the result of the discretized calculation is, inevitably, only approximate, note that, the more specialized a method is, the more accurate it is, and the more effective in terms of computational cost.

The discretization of the problem at hand may take place very early on in the method, as happens with the finite-difference time-domain (FDTD) method, or else at the end of the calculation, as happens with the MoM.

For calculating the radiation of a diffracting structure or antenna, the method of moments is the most suitable tool, but in view of the difficulty of determining the Green's function by prior analysis, another technique is preferable – for example, the FDTD which, today, is the most predominant of all methods.

A frequency-domain method used, when the structures are excessively complex, may be the *finite-element method* (FEM), which consists of solving known differential equations. It is based on a mesh model of the 3D space and a variational formulation of Maxwell's equations. Therefore, it is a robust tool. It is a general method, which has long been used in all areas of physics.

Another is the *finite-difference method*, which, before any operations take place, is based on a 3D mesh. It has the advantage of being relatively simple to implement, taking up less memory. Numerous developments have related to a variant of this method in the time domain, such as FDTD and FVTD. The results obtained in the time domain can be expressed in the frequency domain in a single step, by a Fourier transform. It is then easy to determine the characteristics of the antennas as a function of their frequencies, without needing to carry out a simulation for each frequency point. This represents a major advantage in the study of antennas – particularly for certain antennas with acute resonance.

Another type of digital methods for solving Maxwell's equations is *asymptotic techniques*. These require fundamental approximations of Maxwell's equations – approximations whose validity increases asymptotically with increasing frequency. Examples include physical optics (PO), geometrical optics (GO) and the uniform theory of diffraction (UTD).

As indicated above, asymptotic and numerical methods are respectively classified into two domains (Figure 5.1): High Frequency and Low Frequency – the latter are classed as *exact* methods. The boundary between the two domains is rather fuzzy, with the former enabling the hybridization of the latter. Depending on the dimensions of the problem, and whether the complexity of its implementation is compatible with the computer's power, we can apply one or other, or indeed a hybridization of the two (Figures 5.1 and 5.2). For further details, readers may refer to the following, in particular: Dettman 1962; Booton 1992; Pozar 1998; Chari 2000; Sullivan 2000; Peterson 2001; Volakis 2001; Jin 2002; Tretyakov 2003; Eom 2004; Davidson 2005; Taflove 2005; Gibson 2008; Garg 2008; Moharos 2012; Ufimtsev 2014.

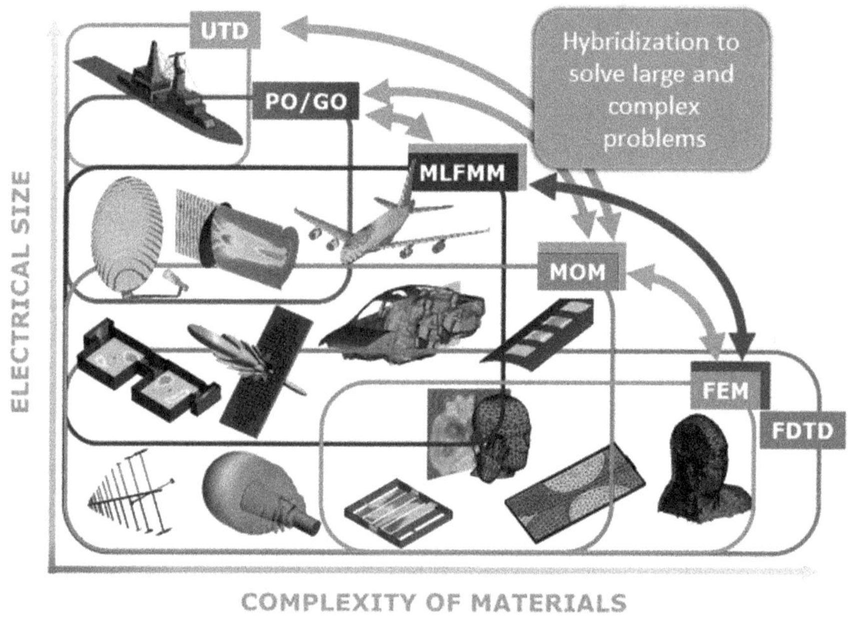

Figure 5.2. *Fields of application of the main digital techniques used. MLFMM = Multilevel Fast Multipole Method (source: www.feko.info)*

The following are a few recent methods developed to optimize and speed up the computations (see the References):

– Higher-Order Techniques as Vector Element, Integral Equation Methods;

– Hybridization with Boundary Integral Methods;

– Hybridization with Asymptotic Methods;

– Fast Multipole Method;

– Multilevel Fast Multipole Method;

– Fast High Order Method.

5.1.2. *Hybridization of digital methods*[2]

Figures 5.1 and 5.2, above, indicate the possibilities for hybridization between certain methods, depending on the problem at hand. The hybridization of digital methods is one of many strategies being pursued in search of speed, efficiency and accuracy of electromagnetic modeling of complex structures, combining electrically long, regularly shaped parts and parts with complex shapes and a shorter electrical length. Instead of a single, all-encompassing formulation, our goal is to apply hybrids comprising multiple digital methods.

5.1.3. *Low-frequency methods*

5.1.3.1. *Introduction to the method of moments (MoM)*[3]

The method of moments is the most suitable tool to determine the radiation of antennas formed by wires or conductive surfaces. It is the most natural method to apply, but it is founded on the study of the Green's functions. Analytical developments go as far as possible, using an integral method. Then, as a final step, the method of moments is used to express the solution in matrix form. It yields results in the frequency domain.

The method of moments is a versatile computational method, so has given rise to numerous derivations, but it is heavily dependent on the Green's function. It can be used to solve differential equations, integral equations and integro-differential equations in the form $L(f) = g$, where L is a linear operator. It is the ideal method to

2 (Hagdahl 2005; Molinet 2005; Garg 2008).

3 (Harrington 1993; Davidson 2005; Gibson 2008; Huang 2008).

use in the field of antennas, because it uses an integral surface equation, and the integration domain is limited to the metal surfaces. The presence of single- or multilayer dielectrics is taken into account by the determination of the Green's function, which may prove highly complex. The conventional method of moments becomes extremely memory-hungry in the case of large antennas (e.g. in the case of networks of antennas) which generates millions of unknowns. Thankfully, this problem is greatly simplified by the development of a variety of solvers such as the Fast Multipole Method, the Adaptative Integral Method and Fast Fourier Transforms (FFT).

In the MoM, the radiating or diffracting structures are replaced by currents across segments or elements of the surface (samples). It is normally surface currents (the volumic currents may be used in the case of non-homogeneous dielectrics, but are costly in terms of computation). The solving method comes at the end of the process, having taken the analytical equations for the problem as far as possible, with the determination of the Green's function. One of the main advantages of this technique lies in the variational nature of its solution. It is ideal for analyses of radiation and coupling. The limitation on its use is the electrical length of the object with respect to the wavelength λ. Depending on the complexity of the object, a solution can be obtained by using a hybridization of the method. The advantage to the MoM is that it is "old", and therefore well-known, and benefits from the developments made over its lifetime. It is used in numerous commercial software tools, including NEC-4, Momentum, Ensemble, Sonnet, IE3D, Saphir and FEKO.

The principle of the method is to determine unknown quantities on the basis of a set of known functions, weighted by unknown coefficients, and make the resulting equation into a system of linear equations by the imposition of boundary conditions. The result of the linear system is solved numerically by inversion of a matrix. Consider the following general problem:

$$L(f) \approx g \qquad [5.1]$$

where L is a linear operator, and f and g are two scalar or vector functions: g a known function, f an unknown function; it is the response to be determined. In physical problems, L represents the system of equations and g the excitation of the system.

To solve the problem, we begin by expressing the unknown solution by breaking it down on a base of functions (f_i) such that:

$$f \approx \sum_{i=1,N} \alpha_i f_i \qquad [5.2]$$

where the a_i values are unknown complex constants for each term in the series, and need to be determined. This step is called discretization of the function *f*. The choice of the functions f_i in the domain of *L* determines the efficiency of the method. The base functions may cover the entire domain or be different for different subdomains. The accuracy of the calculation result may be assessed by the equation $R = L(f) - g$.

Thus, using the linearity of *L*, we have:

$$R = \sum_{i=1,N} \alpha_i L(f_i) - g \qquad [5.3]$$

Expression [5.3] is a simple linear equation with a number *N* of unknowns α_i which can be determined if we simultaneously generate a number *N* of equations of the same type. To do this, we choose a set of *N* test functions (or weighting functions) $w_1, w_2, w_3, \ldots\ldots w_N$ in the domain of *L* by taking the interior product of *L*(*f*) with each w_i. The whole structure being studied is subdivided into small segments and surface portions. The subdivisions are very small in relation to the wavelength. The interior product (scalar product) between two real or complex functions is defined as follows:

$$< \Psi, \Phi >= \iint \bar{\Psi}(\vec{r})\Phi(\vec{r})ds \qquad [5.4]$$

where $\bar{\Psi}(\vec{r})$ is the conjugate of $\Psi(\vec{r})$.

The interior product is also called the *moment*, which gives rise to the name *method of moments*. Formula [5.4] is the projection of the function $\bar{\Psi}(\vec{r})$ onto the function $\Phi(\vec{r})$. By taking the interior product of [5.4] with a weight function (or weighting function) and zeroing it, we have:

$$< R, w_m = 0 \; m = 1,2,3 \;,\ldots \mathrm{M}$$

or, in its expanded form:

$$\sum_{i=1,N} \alpha_i < w_m, L(f_i) >=< w_m, g >$$

$$m = 1, 2, 3, \ldots M \qquad [5.5]$$

The choice of M and the weighted basis functions N is important for the precision of the result, but also to limit the occupation of the computer memory.

NOTE.– The set of equations represented by [5.3] are independent if and only if the test functions are linearly independent.

Consider the matrix representation:

$$[Z][A] = [B] \qquad [5.6]$$

where:

$$[Z] = \begin{bmatrix} < w_1, L(f_1) >, < w_1, L(f_2) >, \ldots\ldots\ldots, < w_1, L(f_N) > \\ < w_2, L(f_1) >, < w_2, L(f_2) >, \ldots\ldots\ldots, < w_2, L(f_N) > \\ \ldots\ldots\ldots\ldots\ldots\ldots\ldots\ldots\ldots\ldots \\ \ldots\ldots\ldots\ldots\ldots\ldots\ldots\ldots\ldots\ldots \\ < w_M, L(f_1) >, < w_M, L(f_2) >, \ldots\ldots\ldots, < w_M, L(f_N) > \end{bmatrix} \ldots \qquad [5.7]$$

$$[A] = \begin{bmatrix} \alpha_1 \\ \alpha_2 \\ \cdot \\ \alpha_N \end{bmatrix} \ldots\ldots \qquad [5.8]$$

$$[B] = \begin{bmatrix} < w_1, g > \\ < w_2, g > \\ \ldots\ldots\ldots \\ \ldots\ldots\ldots \\ < w_M, g > \end{bmatrix} \qquad [5.9]$$

Next, the matrix equation $[Z][A]=[B]$ is solved for the vector $[A]$, using an appropriate technique. Finally, the coefficients α_i are substituted into $f \approx \sum_{i=1,N} \alpha_i f_i$ to determine *f* by means of the basis vector:

$$f=[f]^t\,[A]$$

where:

$$[f]^t=[f_1 \quad f_2 \quad . \quad f_N]$$

The choice of basis functions f_i and test functions w_i (weighting, weight) is key to obtaining an effective, precise solution. In general, basis functions must be as close as possible to the unknown function. There are many possibilities in the choice of basis functions and test functions, but a limited number are used in practice (Gibson 2008). The elements of these functions must be linearly independent. The elements of the matrix [Z] must also be linearly independent in order for the solution to be unique. Similarly, the elements must be chosen in such a way as to minimize computation time. If the basis functions f_i and test functions w_i are the same, this special procedure is known as the *Galerkin method.* Note that rooftop-type basis functions (a rectangular mesh of the structure) do not produce satisfactory results in terms of modeling complex structures that exhibit discontinuities. In order to simplify the resolution of the integrals of the matrix [Z], the analytical technique for calculating them uses RWG basis functions, developed by Rao, Wiltonet and Glisson (Rao *et al.* 1982). These functions are defined across a pair of triangular cells associated with a shared internal edge. They are best suited to the modeling of structures which are arbitrary in shape (Ghannay 2008).

In electromagnetics, the equations solved by the MoM are generally expressed as the electric-field integral equation (EFIE) (Hodges 1997; Jung 2002) or the magnetic field integral equation (MFIE). These two equations are derivatives of Maxwell's equations. The problem is formulated as follows, where *n* is the normal to a conductive surface:

$$\text{EFIE: } \hat{n} \wedge \vec{E}_{incident} = \vec{f}_e\left(\vec{J}_S, \vec{M}_S\right) \qquad [5.10]$$

where $\vec{E}_{incident}$ is the known source; $\vec{J}_S$ and $\vec{M}_S$ are the unknown electric and magnetic surface current densities, which we need to determine. In general, software tools based on the EFIE yield excellent results with open geometries for which the electric field predominates in the zone near to the source.

$$\text{MFIE: } \hat{n} \wedge \vec{H}_{incident} = \vec{f}_m\left(\vec{J}_s, \vec{M}_s\right) \quad [5.11]$$

where $\vec{H}_{incident}$ is also the known source. Software based on the MFIE is a wise choice when dealing with geometries with current circulation which predominates, where the magnetic field is dominant in the near zone.

Depending on the diffracting or radiating structure to be dealt with, we use one or other of the two integral equations (EFIE or MFIE). Note that the magnetic current densities $\vec{M}_S$, and also the magnetic charges ρ_m, have no physical existence, but are used as mathematic tools to solve problems involving radiation and diffraction. The software tools may employ one or other of the two techniques EFIE or MFIE, but at certain frequencies, in the case of cavities, they may prove to be unstable. To remedy this problem, numerous software packages combine the two methods to produce a hybrid method known as the Combined Field Integral Equation (CFIE); it requires more memory to invert the matrix $[Z]$, but yields a more stable result (Ergûl 2006; Hubing 2008).

5.1.3.1.1. *A few of the strengths of the MoM*[4]

In the case of the typical uses made of the MoM, such as radiation of antennas and diffraction:

– it is capable of efficient treatment of the conductive surfaces. Only the surfaces of the object are meshed. No free-space region around the antenna is mapped as a mesh. With wire antennas, the treatment is even more efficient, because the discretization is one-dimensional;

– the MoM includes the condition of infinite radiation, verifying correct behavior of the far field, which is of particular importance in problems involving diffraction;

– the surface currents calculated by the MoM can be used to calculate the parameters of the antenna (impedance, gain, radiation pattern, etc.). By the Sommerfeld potentials, an efficient formulation can be derived for stratified media: single- or multi-layer printed antennas, components, microstrip supply lines;

4 (Davidson 2005; Gibson 2008; Garg 2008).

– it offers the use of very high-performing, stable software programs including NEC-4, SupNEC and FEKO, among others;

– it is attractive for the study of large metal surfaces and homogeneous materials.

5.1.3.1.2. A few of the weaknesses of the MoM

– It does not take account of electromagnetically penetrable materials and differential equation formulas. If the materials are homogeneous, it is possible to use an equivalent surface currents formulation, but for non-homogeneous materials, we must use fictitious equivalent volumic currents, which is extremely costly in terms of computational power.

– Certain formulations of the MoM, particularly those based on the MFIE, require closed surfaces, which is often impractical (Hodges 1997; Jung 2002; Hergûl 2006; Hubing 2008).

– The Green's function is particularly tricky to determine, especially in the case of stratified media.

– Depending on its size, the inversion of the matrix with complex coefficients may require specific solvers.

– Resolution of a problem with the MFIE is not appropriate for open surfaces, unlike the EFIE.

In conclusion, the MoM is an effective method for use in the frequency domain to solve problems of radiation and diffraction in the case of wires or perfectly conductive surfaces. It is not the best choice if the problem involves non-homogeneous dielectric materials. In this case, a hybridization with the FEM may be a good solution.

5.1.3.2. *Introduction to the Finite Difference Time Domain (FDTD) method*[5]

The Finite Difference Time Domain (FDTD) method is a numerical technique used to solve time-domain differential equations on a computer.

The FDTD belongs to the same vintage of methods as the MoM and the FEM. It was put forward in the field of electromagnetics by Kane S. Yee in 1966. Like the FEM, it is based on Maxwell's partial differential equations. It does not involve a Green's function. Like the FEM, the FDTD does not use basis functions or test functions, as the Method of Moments does. It gives a direct approximation of Maxwell's differential operators by a first-order Taylor expansion in time and space.

5 (Yee 1966; Taflove 2005; Davidson 2005).

Since the dawn of wireless telecommunications, it has received great interest and flourishing development, owing to the need to miniaturize handsets, the study of the impact of electromagnetic (radio-frequency – RF) radiation on the human body, the study of *stealth* technologies, lightning protection, etc. The study of these problems, with a view to modeling reality as closely as possible for highly diverse systems, has led to new realistic models for the physical parameters appearing in Maxwell's equations: conductivity σ, permittivity ε and permeability μ. Among other avenues of research, these evolutions enable us to explore the optical properties of metal nanoparticles (Francioso 2014), to study metamaterials and photonic crystals, etc. (Joannopoulos 1995; Sarychev 1996; Lavrinenko 2015). Nanotechnologies have extended the possibilities of the FDTD method (Musa 2014) to frequencies in the optical range (Lavrinenko 2015). FDTD is also applied in the domain of fiber-optics. For EMC studies, given its relative ease of use, combined with rapidly growing computing power, FDTD, married with mesh generators and high-performing graphic software, becomes a veritable 3D simulator. It is in the process of overtaking many frequency-based methods, relegating them to the background.

The FDTD method has an *explicit* approach to finite differences: there are no large matrices to be inverted, no variational calculations (Taflove 1995, 2000). One of the drawbacks to this technique lies in the fact that Maxwell's equations are solved in a calculation domain whose dimensions are necessarily finite. Open-ended problems, of theoretically infinite spatial extent, can nonetheless be solved by hybridization with other methods. The boundary conditions at the edges of the calculation volume, thirty years ago, exhibited a major disadvantage, because they only imperfectly imitated the free space, giving rise to parasitic reflections; similarly, the available computation power used to limit the discretization of the volume. Today, with the scalar, vectorial and parallel supercomputers at our disposal, the size of the problem to be studied is no longer prohibitive, either in terms of computation time or *memory* size.[6] In addition, with the Perfectly Matched Layer (PML) technique developed by Berenger in 1994, the boundary algorithm is less complex and highly efficient, while also using a less thick absorbent layer.

5.1.3.2.1. *The principle of FDTD*

The FTDT method uses the approximation of finite differences centered both in space and in time, for the derivatives appearing in Maxwell's equations, particularly in Ampère's and Faraday's laws. Thus, let us consider the Taylor expansion for a continuously differentiable function $f(x)$ At point x_0 within the interval $\pm\Delta/2$:

6 Top 500: https://www.top500.org/.

$$f\left(x_0+\frac{\Delta}{2}\right)=f(x_0)+\frac{\Delta}{2}.f'(x_0)+\frac{1}{2.!}\left(\frac{\Delta}{2}\right)^2 f''(x_0)+\frac{1}{3.!}\left(\frac{\Delta}{2}\right)^3 f'''(x_0)+.....$$

$$f\left(x_0-\frac{\Delta}{2}\right)=f(x_0)-\frac{\Delta}{2}.f'(x_0)+\frac{1}{2.!}\left(\frac{\Delta}{2}\right)^2 f''(x_0)-\frac{1}{3.!}\left(\frac{\Delta}{2}\right)^3 f'''(x_0)+.....$$

where, at point x_0, $f'(x_0)$ is the first derivative at that point, $f''(x_0)$ is the second derivative, $f'''(x_0)$ the third derivative, etc. (see appendix 1).

By subtracting the second series from the first, and dividing by Δ, we obtain:

$$\frac{f\left(x_0+\frac{\Delta}{2}\right)-f\left(x_0-\frac{\Delta}{2}\right)}{\Delta}=f'(x_0)+\frac{1}{3.!}\left(\frac{\Delta}{2}\right)^2 .f'''(x_0)+.....$$

The left-hand term represents the first derivative of the function at point x_0, plus a term which depends on Δ^2.

Similarly, as can be seen from the expansion into a Taylor series, we have an infinite number of other terms. For these other terms, the next depends on Δ^4, etc. This equation can be rewritten as follows:

$$\frac{d.f(x)}{d.x}\bigg]_{x=x_0}=\frac{f\left(x_0+\frac{\Delta}{2}\right)-f\left(x_0-\frac{\Delta}{2}\right)}{\Delta}+O\left(\Delta^2\right)$$

where $O(\Delta^2)$ represents the second-order error committed and overlooked. If Δ is sufficiently slight, the approximation of the derivative is considered reasonable, and we set:

$$\frac{d.f(x)}{d.x}\bigg]_{x=x_0}\approx\frac{f\left(x_0+\frac{\Delta}{2}\right)-f\left(x_0-\frac{\Delta}{2}\right)}{\Delta}$$

We can see that if the powers of Δ are ignored beyond the second order, the precision is of the second order, meaning that if Δ is decreased by a factor of 10, the error as to the value of the derivative is decreased by a factor of approximately 100.

The precision can be improved by taking an additional term in the Taylor expansion, but in order to do so, additional points must be included around x_0 $(-\frac{3}{2}\Delta, -\frac{1}{2}\Delta, +\frac{3}{2}\Delta)$ for every iteration.

The use of a higher-order central difference is certainly possible in the FDTD method, but the cost is that it becomes extremely complicated due to the involvement of neighboring cells in each increment of time and space (see Figure 5.6).

The FDTD method is based on dual discretization: that is, both temporal and spatial discretization. The 3D space is discretized by a uniform mesh in its basic version, evolving toward a non-constant mesh as a function of the objects being studied (Taflove 1995). A constant step $\Delta_x = \Delta_y = \Delta_z$ does not compromise the validity of the algorithm in comparison to one using inconstant steps, other than that it poses a problem in connecting the steps to one another. The time domain is discretized by a temporal increment Δ_t. Thus, each partial differential (be it temporal or spatial) in the equations showing the evolution of the components of the electric and magnetic fields is replaced by a second-order Taylor expansion. Maxwell's equations are written as:

– Maxwell–Gauss equation:

$$\vec{\nabla}.\vec{E} = 0$$

– Maxwell–Thomson equation:

$$\vec{\nabla}.\vec{H} = 0$$

– Maxwell–Faraday equation:

$$\vec{\nabla} \wedge \vec{E} = -\mu_0.\frac{\partial \vec{H}}{\partial t}$$

– Maxwell–Ampère equation:

$$\vec{\nabla} \wedge \vec{H} = \vec{J} + \varepsilon_0.\varepsilon_r.\frac{\partial \vec{E}}{\partial t}$$

NOTE.– Here, for simplicity's sake, we consider a linear, homogeneous and isotropic (LHI) and non-magnetic medium, which does not detract from the widespread applicability of the principle of the method.

$\vec{E}$ is the electrical field, $\vec{H}$ is the magnetic field, $\vec{J}$ is the current density, μ_0 is the magnetic permeability of a vacuum, ε_0 and ε_r are, respectively, the dielectric permittivity of a vacuum, and the relative permittivity of the medium $\varepsilon_r = \varepsilon / \varepsilon_0$.

In a three-dimensional system of Cartesian coordinates, the Maxwell–Faraday and Maxwell–Ampère equations give us six equations:

$$\mu_0 . \frac{\partial H_x}{\partial t} = \frac{\partial E_y}{\partial z} - \frac{\partial E_z}{\partial y}$$

$$\mu_0 . \frac{\partial H_y}{\partial t} = \frac{\partial E_z}{\partial x} - \frac{\partial E_x}{\partial z}$$

$$\mu_0 . \frac{\partial H_z}{\partial t} = \frac{\partial E_x}{\partial y} - \frac{\partial E_y}{\partial x}$$

$$\varepsilon . \frac{\partial E_x}{\partial t} = \frac{\partial H_z}{\partial y} - \frac{\partial H_y}{\partial z} - j_x$$

$$\varepsilon . \frac{\partial E_y}{\partial t} = \frac{\partial H_x}{\partial z} - \frac{\partial H_z}{\partial x} - j_y$$

$$\varepsilon . \frac{\partial E_z}{\partial t} = \frac{\partial H_y}{\partial x} - \frac{\partial H_x}{\partial y} - j_z$$

These are the six equations for the evolution of the components of the EM fields. We observe that the evolution, over time, of the magnetic field is linked to the spatial variation of the electrical field, and vice versa.

5.1.3.2.2. The principle of Yee's algorithm[7]

The system proposed by Yee overcomes the difficulty due to the mutual interdependence of the electric field $\vec{E}$ and magnetic field $\vec{H}$. Indeed, it gives us an explicit numerical diagram which allows us to calculate the electromagnetic (EM) field for the whole volume studied, as a function of time. The components of the EM field are shifted by half a spatial step (spatial offset) and calculated by an alternative

7 (Yee 1966).

and iterative manner, respectively to odd and even multiples of the half time step temporal offset. In 1975, Yee's scheme was generalized by Taflove *et al.* (2000).

5.1.3.2.3. Discretization of Maxwell's equations

Spatial discretization

To model this system of equations, we construct a regular mesh to represent the volume, containing the object we wish to study in all three spatial dimensions $< x,y,z >$, with constant increments $\Delta_x = \Delta_y = \Delta_z$.

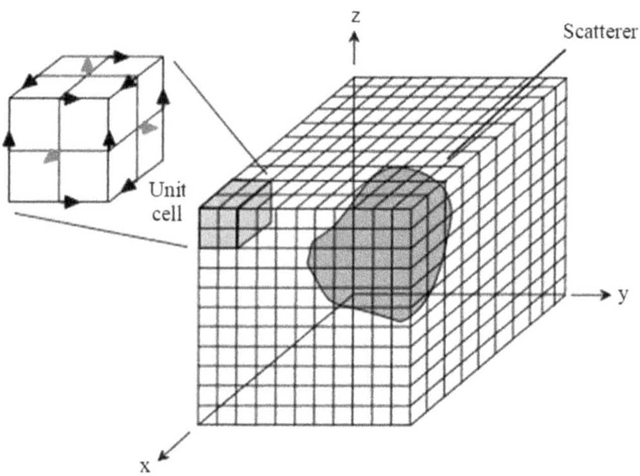

Figure 5.3. *Spatial discretization of the system to be modeled (FDTD)*

The calculation volume is a parallelepiped comprising (n_x, n_y, n_z) cubic cells, signposted by their integer indices *i*, *j* and *k*. These indices, respectively, range from 1 to n_x, from 1 to n_y and from 1 to n_z.

From the first of the six equations, we know that the calculation of H_x involves the partial differential of E_y with respect to *z*, and the partial differential of E_z with respect to *y*. The centered approximation of the first derivative means that the point where we calculate H_x must be situated, at once:

– at the middle of a segment parallel to the *oz* axis, the two ends of which are points where E_y is known;

– at the middle of a segment parallel to the *oy* axis, the two ends of which are points where E_z is known.

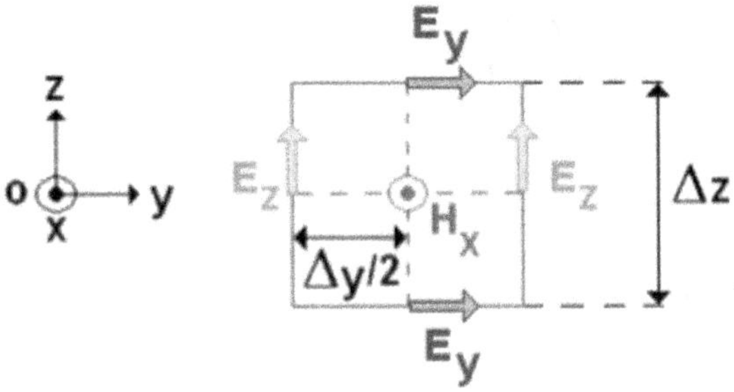

Figure 5.4. *Circulation of the field $\vec{E}$ around the field $\vec{H}$*

Similarly, the point at which we calculate E_x is situated simultaneously:

– at the middle of a segment parallel to the *oz* axis, the two ends of which are points where H_y is known;

– at the middle of a segment parallel to the *oy* axis, the two ends of which are points where H_z is known.

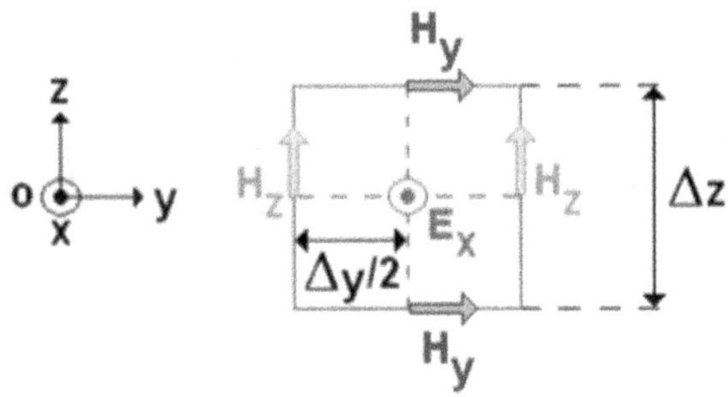

Figure 5.5. *Circulation of the field $\vec{H}$ around the field $\vec{E}$*

The centered approximation of the first derivative applied to the six equations gives conditions which, taken as a whole, produce Yee's system (Yee 1966), represented by Figure 5.6.

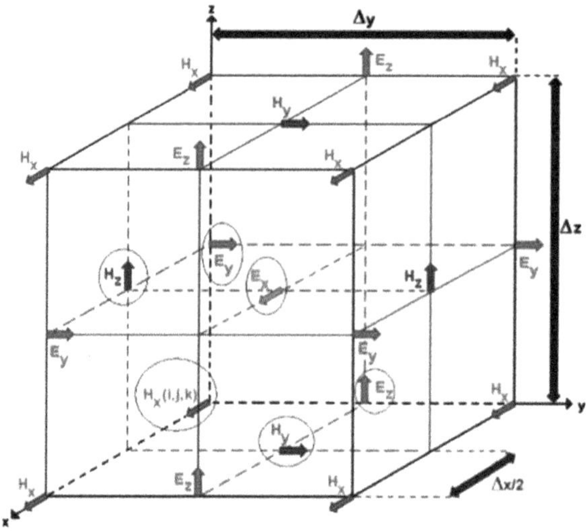

Figure 5.6. *Position of the fields $\vec{H}$ and $\vec{E}$ in Yee's system (Francioso 2014)*

Temporal discretization

On its left-hand side, the first of the six equations includes the temporal derivative of H_x; the right-hand side is considered at time *t*. The principle of the centered derivative means that the components E_y and E_z must be calculated between two successive times at which we calculate H_x. Thus, taking account of the other five equations, we conclude that the electrical field and the magnetic field cannot be calculated at the same time, but rather at offset times. Setting Δt as the time sampling step, we calculate the electrical field for multiple integers of Δt and the magnetic field at half integers of Δt (Figure 5.7).

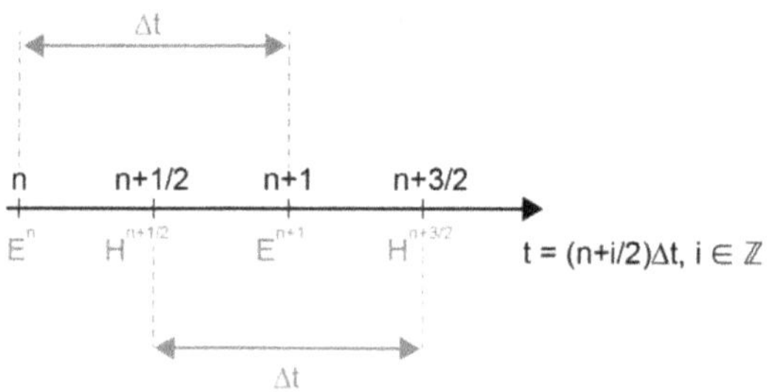

Figure 5.7. *Representation of temporal discretization*

Discretized Maxwell's equations

After discretization, the six partial differential equations respectively become:

$$H_x^{n+\frac{1}{2}}(i,j,k) = H_x^{n-\frac{1}{2}}(i,j,k)$$

$$+\frac{\Delta t}{\mu_0}\cdot\left(\frac{E_y^n\left(i,j,k+\frac{1}{2}\right)-E_y^n\left(i,j,k-\frac{1}{2}\right)}{\Delta z}-\frac{E_z^n\left(i,j+\frac{1}{2},k\right)-E_z^n\left(i,j-\frac{1}{2},k\right)}{\Delta y}\right)$$

$$H_y^{n+\frac{1}{2}}\left(i+\frac{1}{2},j+\frac{1}{2},k\right) = H_y^{n-\frac{1}{2}}\left(i+\frac{1}{2},j+\frac{1}{2},k\right)+$$

$$\frac{\Delta t}{\mu_0}\cdot\left(\frac{E_z^n\left(i+1,j+\frac{1}{2},k\right)-E_z^n\left(i,j+\frac{1}{2},k\right)}{\Delta x}-\frac{E_x^n\left(i+\frac{1}{2},j+\frac{1}{2},k+\frac{1}{2}\right)-E_x^n\left(i+\frac{1}{2},j+\frac{1}{2},k-\frac{1}{2}\right)}{\Delta z}\right)$$

$$H_z^{n+\frac{1}{2}}\left(i+\frac{1}{2},j,k+\frac{1}{2}\right) = H_z^{n-\frac{1}{2}}\left(i+\frac{1}{2},j,k+\frac{1}{2}\right)+$$

$$\frac{\Delta t}{\mu_0}\cdot\left(\frac{E_x^n\left(i+\frac{1}{2},j+\frac{1}{2},k+\frac{1}{2}\right)-E_x^n\left(i+\frac{1}{2},j-\frac{1}{2},k+\frac{1}{2}\right)}{\Delta y}-\frac{E_y^n\left(i+1,j,k+\frac{1}{2}\right)-E_y^n\left(i,j,k+\frac{1}{2}\right)}{\Delta x}\right)$$

$$E_x^{n+\frac{1}{2}}\left(i+\frac{1}{2},j+\frac{1}{2},k+\frac{1}{2}\right) = E_x^n\left(i+\frac{1}{2},j+\frac{1}{2},k+\frac{1}{2}\right)+\frac{\Delta t}{\varepsilon_0\varepsilon_r\left(i+\frac{1}{2},j+\frac{1}{2},k+\frac{1}{2}\right)}\cdot$$

$$\left(\frac{H_z^{n+\frac{1}{2}}\left(i+\frac{1}{2},j+1,k+\frac{1}{2}\right)-H_z^{n+\frac{1}{2}}\left(i+\frac{1}{2},j,k+\frac{1}{2}\right)}{\Delta_y}-\frac{H_y^{n+\frac{1}{2}}\left(i+\frac{1}{2},j+\frac{1}{2},k+1\right)-H_y^{n+\frac{1}{2}}\left(i+\frac{1}{2},j+\frac{1}{2},k\right)}{\Delta_z}-J_x^{n+\frac{1}{2}}\left(i+\frac{1}{2},j+\frac{1}{2},k+\frac{1}{2}\right)\right)$$

$$E_y^{n+1}\left(i,j,k+\frac{1}{2}\right) = E_y^n\left(i,j,k+\frac{1}{2}\right)+\frac{\Delta t}{\varepsilon_0\varepsilon_r\left(i,j,k+\frac{1}{2}\right)}\cdot$$

$$\left(\frac{H_x^{n+\frac{1}{2}}(i,j,k+1) - H_x^{n+\frac{1}{2}}(i,j,k)}{\Delta_z} - \frac{H_z^{n+\frac{1}{2}}\left(i+\frac{1}{2},j,k+\frac{1}{2}\right) - H_z^{n+\frac{1}{2}}\left(i-\frac{1}{2},j,k+\frac{1}{2}\right)}{\Delta_x} - J_y^{n+\frac{1}{2}}\left(i,j,k+\frac{1}{2}\right) \right)$$

$$E_z^{n+1}\left(i,j+\frac{1}{2},k\right) = E_z^{n}\left(i,j+\frac{1}{2},k\right) + \frac{\Delta t}{\varepsilon_0 \varepsilon_r\left(i,j+\frac{1}{2},k\right)} \cdot$$

$$\left(\frac{H_y^{n+\frac{1}{2}}\left(i+\frac{1}{2},j+1,k\right) - H_y^{n+\frac{1}{2}}\left(i-\frac{1}{2},j+\frac{1}{2},k\right)}{\Delta_x} - \frac{H_x^{n+\frac{1}{2}}(i,j+1,k) - H_x^{n+\frac{1}{2}}(i,j,k)}{\Delta_y} - J_z^{n+\frac{1}{2}}\left(i,j+\frac{1}{2},k\right) \right)$$

The above equations are the EM field updating equations. For example, the notation $E_{z^{n+1}}\left(i,j+\dfrac{1}{2},k\right)$ represents the value assumed by the component of the electrical field along the z axis, at the point with coordinates $(x = i.\Delta x, y = (j+\dfrac{1}{2}).\Delta y, z = k.\Delta z)$, and at time $t=(n+1).\Delta t$. The equation for the component E_z shows that at time $t=(n+1).\Delta t$, the component E_z is calculated on the basis of:

– the component E_z at the same point in the discretized 3D space, but at the previous time $(t = n.\Delta t)$;

– the values at time $t=\left(n+\dfrac{1}{2}\right).\Delta t$ of the nearest four surrounding components of the magnetic field.

The six components of the EM field are updated at each moment of time sampling. The evolution of the EM field over time is thus deduced by an iterative calculation, known as a *leapfrog diagram*.

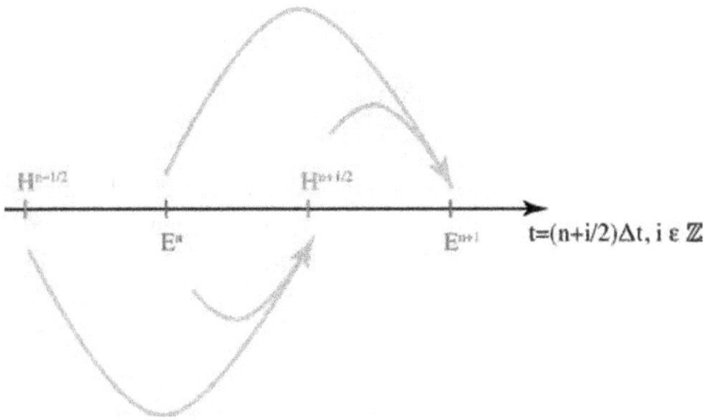

Figure 5.8. *So-called leapfrog diagram (iterative)*

This way of calculating the EM field is intuitive and closely reflects the physical reality, in that a variation of the electrical field engenders a variation in the magnetic field which, in turn, engenders a variation in the electrical field, and so on. The FDTD algorithm can be represented as in Figure 5.9.

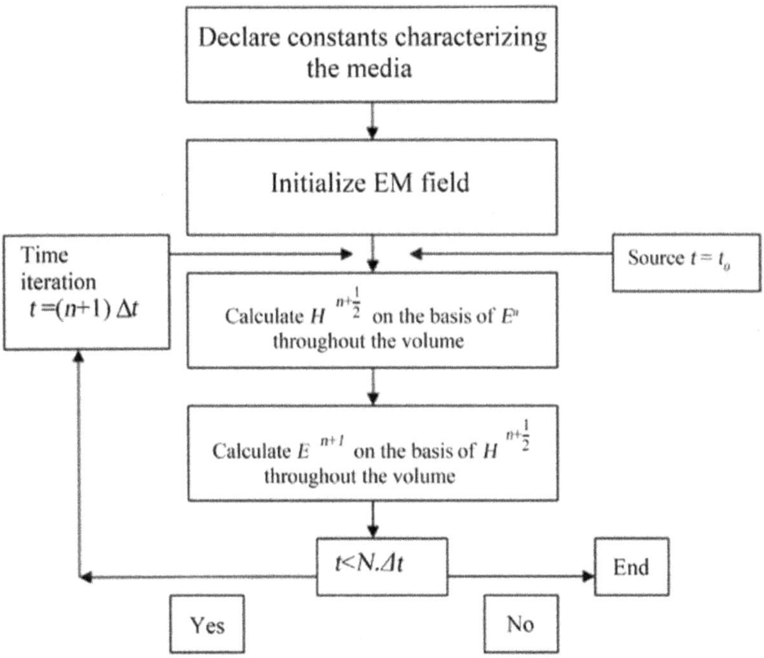

Figure 5.9. *FDTD algorithm*

Choice of spatial and temporal intervals

The choice of the spatial and temporal intervals is not arbitrary. It has an impact on both the accuracy and stability of the FDTD algorithm.

To begin with, the spatial discretization must be sufficiently fine to describe the geometry of the objects contained in the calculation volume as *finely* as possible, and also the distance between the objects making up the structure being analyzed. In addition, the transition from a physical problem in continuous timespace to a sampled timespace creates a parasitic effect known as *numerical dispersion*. This effect arises from the errors made in evaluating the speed of wave propagation in the calculation domain (in a cell, the path is longer along the diagonals). To minimize this effect, the value of the spatial steps is fixed as a function of the wavelength of the highest frequency to be analyzed (Taflove 2000; Tarricone 2004):

$$Max(\Delta x, \Delta y, \Delta z) \leq \frac{\lambda_{\min}}{10}$$

From this equation, it becomes apparent that it is necessary to have at least ten cells per wavelength λ to obtain correct results.

In addition, temporal iterative algorithms such as the FDTD may give rise to incorrect inflation – usually exponential inflation – of the values of the EM field, leading the calculation to diverge from the true solution. To combat this problem, when choosing the time increment, we must take care to respect a criterion known as the *numerical stability criterion* or the *Courant–Friedrichs–Lewy condition* (Courant 1967), which supersedes Yee's condition (Yee 1996; Taflove 1975), and is written as:

$$\Delta t = \alpha.\Delta t_1 \quad \alpha \in]0,1]$$

with:

$$\Delta t_1 \leq \frac{1}{v\sqrt{\dfrac{1}{(\Delta x)^2} + \dfrac{1}{(\Delta y)^2} + \dfrac{1}{(\Delta z)^2}}}$$

where v is the velocity of the EM wave in the medium with constant ε_r and μ.

This stability condition means that the numerical sampling must be sufficiently fine-grained to follow the evolution of the EM field over time. In other words, the speed of the numerical algorithm must be greater than the velocity of the wave propagation, so:

$$v < \frac{\sqrt{(\Delta x)^2 + (\Delta y)^2 + (\Delta z)^2}}{\Delta t} = \text{speed of algorithm}$$

If that stability condition is not adhered to, the algorithm will diverge.

Case of imperfect, dispersive materials

To consider dielectric or metal materials to be perfect is an approximation (see section 2.2). For example, it is impossible to have a lossless dielectric outside of free space. In the case of free space, the losses are deemed to be negligible. The dielectric permittivity ε and permeability μ, respectively, describe the electrical and magnetic properties of the materials. In the case of dispersive materials, the dielectric permittivity and magnetic permeability depend on the frequency: $\varepsilon = \varepsilon(\omega)$, $\mu = \mu(\omega)$. They may also depend on the position within the material. The dielectric permittivity reflects the phenomena of polarization induced in the material with the propagation of an EM wave (Kittel 2007).

As the FDTD is a temporal method, Maxwell's equations, in which the frequency appears explicitly in the dielectric and/or magnetic function, cannot be implanted. There are various methods available to solve this problem (Francioso 2014; Pereda 2002; Liu 2003; Vial 2005; Schneider 2017).

Multiplication of harmonic functions is equivalent to a convolution in the time domain. However, it is not easy to switch from the frequency domain to the time domain. From the expressions of the fields $\vec{E}$ and $\vec{H}$, we have seen that:

$$\vec{D} = \varepsilon_0 \vec{E} + \vec{P}$$

and:

$$\vec{H} = \frac{\vec{B}}{\mu_0} - \vec{M}$$

where $\vec{P}$ and $\vec{M}$, respectively, represent the electrical and magnetic dipoles induced in the medium subjected to an EM wave. At a given frequency, for a linear and isotropic medium, the polarization vectors can be written thus (Al-Jabr 2009; Schneider 2017):

$$\vec{P}(\omega) = \varepsilon_0 . \chi_e(\omega) . \vec{E}(\omega)$$
$$M(\omega) = \chi_m(\omega) . \vec{H}(\omega)$$

where $\chi_e(\omega)$ and $\chi_m(\omega)$ are, respectively, the electrical and magnetic susceptibilities of the materials used. They represent the response of these materials as a function of the frequency. These susceptibilities have different expressions depending on the Drudge or Lorentz "linear" medium, or "nonlinear" for the Kerr effect (Ammann 2007). We can write:

$$\vec{D}(\omega) = \varepsilon_0 . \vec{E}(\omega) + \varepsilon_0 . \chi_e(\omega) . \vec{E}(\omega)$$

$$\vec{B}(\omega) = \mu_0 . \vec{H}(\omega) + \mu_0 . \chi_m(\omega) . \vec{H}(\omega)$$

Consider the expression of the permittivity $\varepsilon(\omega) = \varepsilon_0(\varepsilon_\infty + \chi_e(\omega))$, where the constant ε_∞ takes account of the effect of the charged medium at a high frequency where the susceptibility tends toward zero. Similarly, consider the expression of the permeability $\mu(\omega) = \mu_0(\mu_\infty + \chi_m(\omega))$ where μ_∞ takes account of the permeability at high frequency.

After a Fourier transform and a convolution product to switch from the frequency domain to the time domain, an implantation of $\varepsilon(t)$ and/or $\mu(t)$ is possible with the Auxiliary Differential Equation (ADE) approach. It consists, on the basis of the displacement vector make single full-size space $\vec{D}(\omega)$ in the frequency domain, of:

$$\vec{D}(\omega) = \varepsilon_0 \varepsilon_r(\omega) \vec{E}(\omega)$$

In the equation $\vec{D} = \varepsilon_0 \vec{E} + \vec{P}$, the polarization vector is replaced by a sum M of $\vec{P}_n$ partial densities of electric dipole moments. The expression of the $\vec{P}_n$ values depends on the Drude or Lorentz model, or the Debye model (Al-Jabr 2009; Ammann 2007; Francioso 2014; Schneider 2017), for materials considered:

$$\vec{D}(\omega) = \varepsilon_0 \varepsilon_\infty \vec{E}(\omega) + \varepsilon_0 \sum_{m=0}^{M} \vec{P}_m(\omega)$$

We move into the time domain by means of an inverse Fourier transform:

$$\vec{D}(t) = \frac{1}{2\pi} \int_{-\infty}^{+\infty} \vec{D}(\omega) e^{+j\omega t} d\omega$$

The ADE-FDTD algorithm introduces 2(M+1) new differential equations to update the components of the electrical field in dispersive materials. The equations to update the magnetic field remain unchanged. One possible ADE-FDTD algorithm is shown in Figure 5.10 (Francioso 2014).

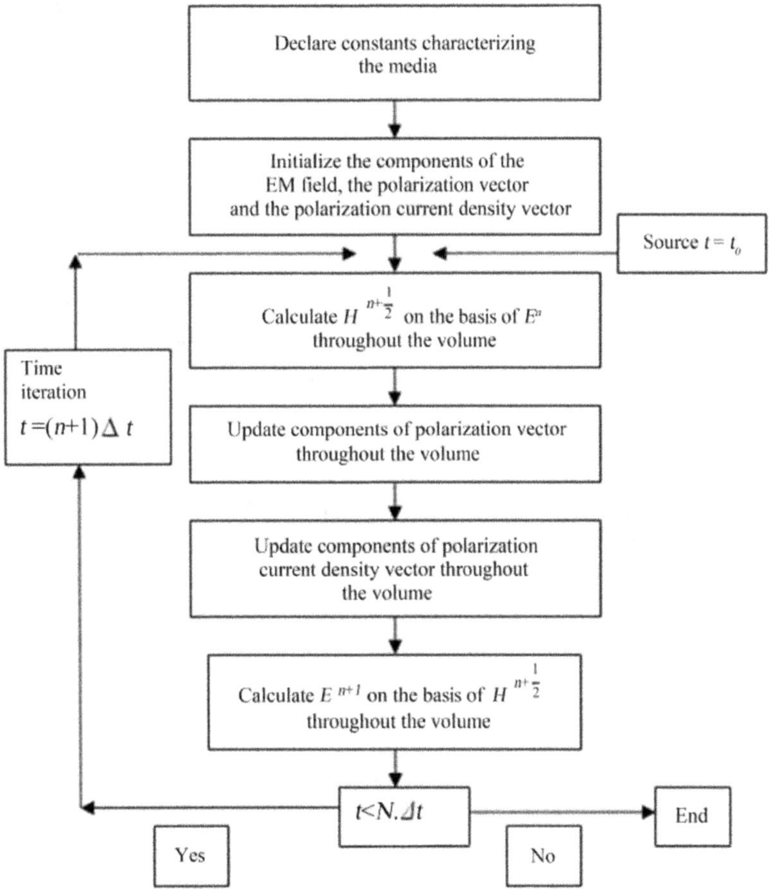

Figure 5.10. *Example of an ADE-FDTD algorithm (Francioso 2014)*

Truncation of the calculation domain with PMLs (Perfectly Matched Layers)

The calculation volume is finite (Figure 5.3). Modeling an open physical system where the waves can propagate infinitely as a discretized and bounded 3D space, poses the problem of parasitic reflections on the edges of the calculation volume. The equations of the FDTD, therefore, cannot be used just as they are, but must be altered in order to respect particular boundary conditions on the calculation domain. Today, the best performing and most widely used model is the Perfectly Matched Layers (PML) model and its derivatives (Bérenger 1994, 1996). Bérenger showed that it was possible to simulate the free space by adding a layer, all around the calculation domain, composed of an imaginary material absorbing non-null electrical conductivity σ and magnetic conductivity σ. Perfect matching between two media $(\varepsilon_1, \mu_1, \sigma_1, \sigma_1)$ and $(\varepsilon_2, \mu_2, \sigma_2, \sigma_2)$ is achieved by satisfying the four conditions which are governed by the system below:

$$\varepsilon_1 = \varepsilon_2$$

$$\mu_1 = \mu_2$$

$$\frac{\sigma_1}{\varepsilon_1} = \frac{\sigma_1^*}{\mu_1}$$

$$\frac{\sigma_2}{\varepsilon_2} = \frac{\sigma^*{}_2}{\mu_2}$$

5.1.3.2.4. A few strengths of the FDTD[8]

The code is exceptionally simple to develop in its fundamental version for a *full-wave* code, which requires less work than the MoM or the FEM, or indeed other methods. This is tempered by the fact that certain details of the FDTD method may take some time to implement, even in the case of problems which seem simple to begin with but prove not to be as the calculation proceeds. Nevertheless, it is the only method that can be adapted to suit a specific problem within a reasonable time.

– Materials exhibiting inhomogeneities are easy to deal with using FDTD, as is the case with the FEM.

– It has the capacity for fairly precise geometric modeling (but is not as universally applicable as the FEM due to the staircasing effect on the mesh on non-orthogonal grids). Commercial codes often include extensions to the method to rectify this, so it is not necessarily a problem.

8 (Davidson 2005).

– As the method is in the time domain, the frequency response across a wide range can be obtained in one sweep, using a Fourier transform.

– It is particularly effective to limit the calculation volume by using PMLs.

– To model a new structure, the only problem is the generation of a new mesh, which is simpler than the difficult task of reformulating an integral equation.

– The method does not use matrix equations, so there is no matrix inversion.

– It can easily be hybridized, on the basis of the equivalence theorem.

5.1.3.2.5. A few weaknesses of the FDTD

– The switch from an orthogonal mesh to a non-orthogonal or matching mesh is complex, which lessens the attractiveness of the method.

– The boundary conditions on dielectric or conductive objects are imprecise, due to the offsetting of a semi-grid of the electrical field and the magnetic field.

– Media with loss require a great deal of effort in setting up the algorithm, but very good results are obtained, from microwave frequencies to optical frequencies, for materials as complex as metamaterials.

Like the FEM, the FDTD method is not as effective as the MoM in problems involving radiation and diffraction for entirely metallic, perfectly conductive materials.

5.1.3.3. *Introduction to the finite element method (FEM)*[9]

The FEM is a frequency-domain method. It is used when the structures are too complex to be efficiently processed by the method of moments. It is based on a mesh of the 3D space and on a variational formulation of Maxwell's equations; therefore, it is highly robust. First put forward in the field of electromagnetics by Silvester (1969), the finite element method (FEM) is of the same vintage as the MoM and the FDTD method. It is used to determine an approximate solution on a spatial domain, meaning that it allows us to calculate a field (scalars, vectors, tensors) which corresponds to certain equations and certain set conditions.

This method was initially developed for engineering, and it is in that area that it performs best – particularly electromagnetic and antenna engineering. However, the FEM is not as popular as the method of moments and FDTD because its form is more complicated, in terms of volumic generation of its mesh. On the other hand, it

9 (Silvester 1969, 1990; Volakis 1998; Picon 2009; Jin 2014).

is unparalleled in terms of modeling highly complex structures and materials, using curvilinear, triangular or tetrahedral elements. The FEM, in the time domain, can be formulated as unconditionally stable. Unconditional stability is critical for the analysis of complex antennas. Like FDTD, the FEM solves partial differential equations directly, without resorting to Green's functions. Like FDTD, it requires a three-dimensional discretization of the volume surrounding the antenna. The truncation of that volume is the subject of a great deal of research.

The FEM is based on an interpolation for each element. The method consists of dividing the spatial domain into small elements, also known as meshes, and seeking a simplified formulation for the problem on each element – that is to say, to transform any system of equations into a system of linear equations. Each system of equations can be represented by a matrix. The systems of equations for all the elements are then brought together, forming a large matrix. By solving this global system, we obtain the approximate solution to the problem. Like all numerical methods, it requires a powerful computer. Modern tools like the FEM and other methods have a graphic interface which makes them relatively simple to use. In addition, a certain number of tasks can be automated.

Generally speaking, a given problem can be solved using the FEM in four steps (Jin 2014):

1) *subdivision of the entire domain*: the first step is to subdivide the entire domain into multiple sub-domains or elements. For a regular structure, rectangles are chosen. For a complex shape, triangles are the preferred elements. For reasons of implementation, information must be provided about the nodes and edges of each element, such as the coordinates of the nodes, the number of edges throughout the structure (i.e. the overall numbers), the edge numbers in the particular element (local numbers), etc.;

2) *definition of the interpolation functions*: the second step is to choose the interpolation functions which can be used to approximate the unknown parameters (in our case, the electric fields) in each element. The interpolation functions are often polynomials of order 1, 2, etc., which are defined solely in the element in question;

3) *formulation of the system of equations* – this phase includes three steps:

- formulation of the elementary equations in each element, using either the Ritz or Galerkin method,

- summation of the elementary equations in all of the elements to form a system of equations; this process is known as assemblage,

- imposition of the boundary conditions to obtain the final system of equations;

4) *solution of the system of equations using direct or iterative methods.*

As with the MoM, at the end, the problem is also converted into a matrix equation.

A finite element model may contain a hundred thousand, or even over a million elements. To solve it, therefore, is phenomenally costly in terms of both resources and time: many hours, or even days of computations, which may require parallel processes to be launched on a supercomputer. Thus, the use of high-performing computational algorithms is not inherent to the finite element method, but is nonetheless essential for reasons of performance.

To simplify the principle, consider an example in 2D space (Cardoso 2017). Figure 5.11 shows a domain discretized into finite first-order triangular elements. Consider a scalar function with two variables $F(x,y)$ which represents, for example, a potential with known value $F(x_i,y_i)$ at certain points (x_i,y_j) in the total definition set. This function is approximated by a function $f(x,y)$ which takes the values $F(x,y)$ at the points (x_i,y_j). $f(x,y)$, and is known as the nodal function of $F(x,y)$.

We can define the error $e(x,y)$ of this approximation as:

$$e(x,y)=F(x,y)-f(x,y)$$

The approximation error is zero for the points (x_i,y_j) in the elements of the mesh of the structure under examination.

Figure 5.12 shows, in relief, a surface corresponding to the exact solution of the function $f(x,y)$ and the surface corresponding to its approximation formulated by its values at the nodes, and interpolated by planes connected by the values at the three nodes of an element (Konrad 1973; Silvester 1978; Cardoso 2017).

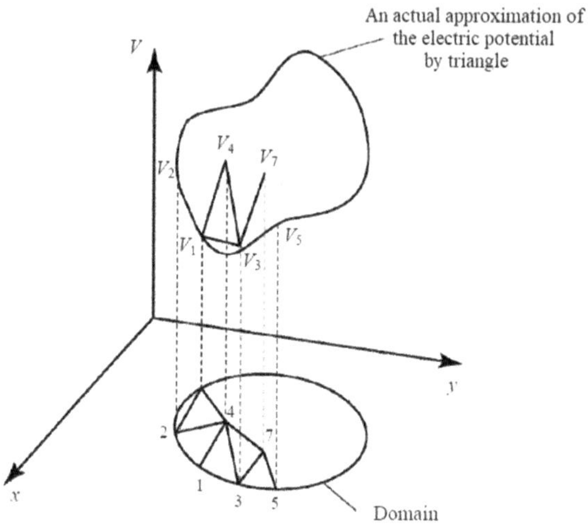

Figure 5.11. *2D linear interpolation*[10]

Figure 5.13 represents a 2D linear interpolation applied to a generic triangular element, extracted from a finite element mesh with two linear polarizations (Orange Lab in La Turbie).

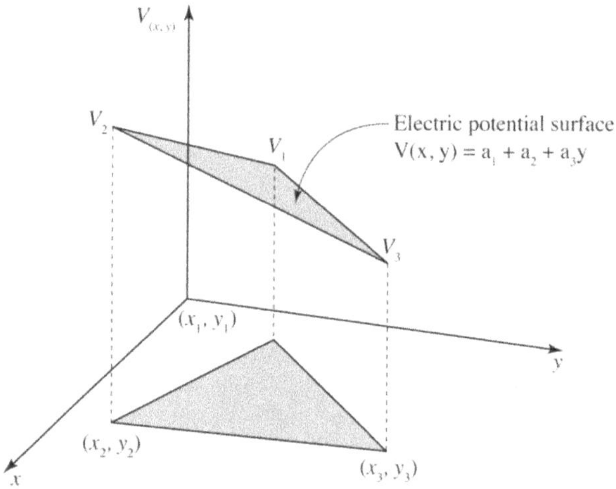

Figure 5.12. *Linear interpolation for a triangular element*

10 (Picon 2009; Cardoso 2017).

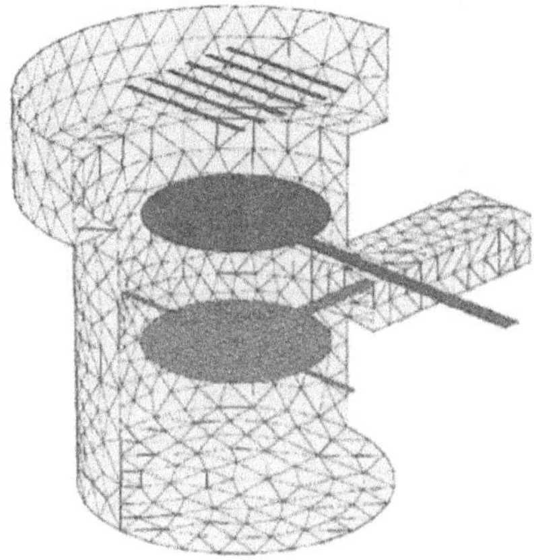

Figure 5.13. *Example of a mesh for an elementary source in the Ku band*

The aim now is to define an interpolation function which gives an *estimation* of the value of the function $f_i = f(x_i, y_i)$ within each element of the discretized object, if its exact value is known at the nodes.

By simplification of the demonstration of the FEM, the interpolation function can be taken in the form:

$$f(x,y) = \alpha_1 + \alpha_2 x + \alpha_3 y$$

where the coefficients α_1, α_2 and α_3 need to be determined.

At each of the three nodes, we have:

$$f_1 = \alpha_1 + \alpha_2 x_1 + \alpha_3 y_1$$

$$f_2 = \alpha_1 + \alpha_2 x_2 + \alpha_3 y_2$$

$$f_3 = \alpha_1 + \alpha_2 x_3 + \alpha_3 y_3$$

so:

$$\alpha_1 = \frac{1}{2\Delta}(a_1 f_1 + a_2 f_2 + a_3 f_3)$$

$$\alpha_2 = \frac{1}{2\Delta}(b_1 f_1 + b_2 f_2 + b_3 f_3)$$

$$\alpha_3 = \frac{1}{2\Delta}(c_1 f_1 + c_2 f_2 + c_3 f_3)$$

with:

$$a_1 = x_2 y_3 - x_3 y_2 \; ; \; b_1 = y_2 - y_3 \text{ , etc., and } \Delta = \frac{1}{2}(b_1 c_2 - b_2 c_1)$$

Δ is the surface area of the triangular element.

By replacing these coefficients in the equation $f(x,y) = \alpha_1 + \alpha_2 x + \alpha_3 y$, the nodal approximation function is written:

$$f(x,y) = \sum_{i=1}^{3} N_i(x,y) f_i$$

The f_i values are the nodal variables with $N_i(x,y) = \frac{1}{2\Delta}(a_i + b_i x + c_i y)$ $i=1, 2, 3.$

The functions $N_i(x,y)$ are known as *form functions* or *interpolation functions*. They are such that:

$$N_i(x_i, y_j) = \delta_{ij} \begin{Bmatrix} 1 \text{.......} si \ldots i = j \\ 0 \text{......;} si \ldots i \neq j \end{Bmatrix} \text{ and } \sum_{i=1}^{3} N_i(x,y) = 1$$

where δ_{ij} is the Kronecker delta.

From this, we conclude that the ith form function takes a value of 1 at node *I* and of 0 at the other two nodes.

Note that in the one-dimensional case, the elements of the mesh are segments. We apply the interpolation function to each element. This form has the benefit of ensuring continuity of the function throughout a definition interval.

If we define a segment between two neighboring nodes, the interpolation is simply linear along the element. We then speak of a first-order finite element. If we define a segment containing three nodes, the interpolation is quadratic, and we speak of second-order finite elements. The nodal approximation only involves nodal variables situated on the element and on its boundary.

5.1.3.3.1. Problem of electromagnetics in general[11]

The principle of the method in two-dimensional (2D) space can be generalized to apply to 3D space, but it is tricky to transition from a scalar function $f(x,y,z)$ to a vector function where $\vec{f}(x,y,z)$ represents an electrical field $\vec{E}(x,y)$ or a magnetic field. The fundamental elements are no longer triangles, as represented in Figure 5.13, but tetrahedral base elements, as represented in Figure 5.14.

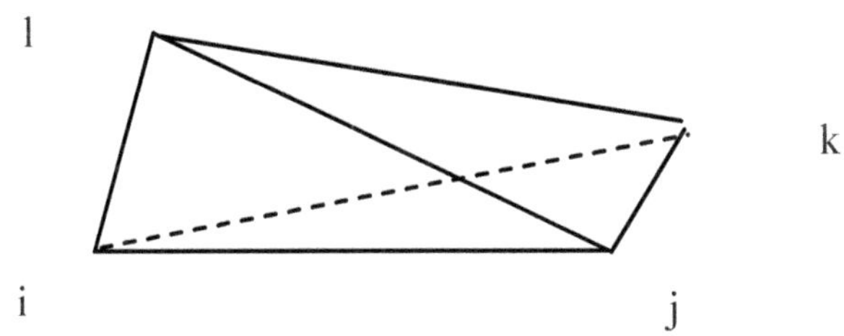

Figure 5.14. *Tetrahedral base element serving to construct a 3D mesh*

The more general case of interest to us here is that of calculating the electromagnetic field in a given calculation volume. We can see that it is sufficient to calculate either the electrical field or the magnetic field.

The differential equation satisfied by one or other of the fields is the Helmholtz equation:

$$\Delta\vec{E} + k^2\vec{E} = \vec{0}$$

$$\Delta\vec{H} + k^2\vec{H} = \vec{0}$$

where $k^2 = \varepsilon_0\mu_0\omega^2$ is the wavenumber in a vacuum and $\Delta = \nabla^2 = \frac{\partial^2}{\partial x^2} + \frac{\partial^2}{\partial y^2} + \frac{\partial^2}{\partial z^2}$.

11 (Picon 2009).

Boundary conditions are imposed at the edges of the calculation volume. Different types of conditions may be imposed:

– *the electric wall condition*: $\vec{E} \wedge \hat{n} = \vec{0}$ or $\vec{H}.\hat{n} = 0$ with $\hat{n}$ being the external normal to the surface;

– *the magnetic wall condition*, used in certain cases of symmetry: $\vec{H} \wedge \hat{n} = \vec{0}$ or $\vec{E}.\hat{n} = 0$.

When the metal is not perfect, we set a surface condition of impedance.

The physical problem is posed on the basis of the differential equation and the boundary conditions. The next step is to construct a functional. Depending on the choice of variable (electrical field or magnetic field), one of the two functionals can be constructed:

$$F\left(\vec{E}\right) = \iiint_V \left(\left(ro\vec{t}\vec{E} \right)^2 - k^2 \varepsilon_r \mu_r \vec{E}.\vec{E} \right) dv$$

$$F\left(\vec{E}\right) = \iiint_V \left(\left(ro\vec{t}\vec{H} \right)^2 - k^2 \varepsilon_r \mu_r \vec{H}.\vec{H} \right) dv$$

In practice, for the subject of interest to us here, the media used are non-magnetic ($\mu_r = 1$). The value of the field for which the functional reaches an extremum provides the solution. To find the field, we break it down into each of the elements, using approximation functions. Unfortunately, the scalar interpolation functions, presented earlier for use with a scalar property such as the potential, are not appropriate to describe the electrical field or magnetic field because, in certain cases, they produce aberrant non-physical solutions, known as parasitic solutions. Numerous explanations for this phenomenon have been given. One is that the interpolation functions presented earlier are not capable of taking account of vectorial properties. In particular, a vector such as the electrical field has zero divergence. However, it is impossible to take account of this constraint simply by using scalar approximation functions, unless we use a penalty term, which makes the method far more unwieldy. A more elegant method is to use vectorial approximation functions, also known as *Whitney forms*. With the scalar approximation functions, we could write:

$$\vec{E}(x,y,z) = \sum_{i=1}^{4} \vec{E}_i N_i(x,y,z)$$

with $\vec{E}_i$ being the value of the field at the interpolation node and $N_i(x,y,z)$ the form functions.

With the Whitney forms, we write:

$$\vec{E}(x,y,z) = \sum_S e_S \vec{\omega}_S(x,y,z)$$

where e_S represents the values of the circulation of the field along an edge and $\vec{\omega}_S(x,y,z)$ the Whitney forms, which are defined by:

$$\vec{\omega}_S = N_j(x,y,z)\,\overline{grad}(N_i(x,y,z) - N_i(x,y,z)\,\overline{grad}\left(N_j(x,y,z)\right)$$

where S is an integer that characterizes an edge situated between nodes i and j. Thus, there are six approximation functions per tetrahedron for so-called *1st-order finite elements*.

Inserting this type of element into the expression of the functional, after differentiation, yields a linear system whose inversion gives the circulations of the fields along the edges. Note that in the case of scalars with the FEM, the unknowns are the three orthogonal components of the field at the nodes of each element. In the case of vectors with the FEM, the unknowns are on the edges of each element.

5.1.3.3.2. Formulation of the problem for antennas

Finite elements are used to simulate antennas. The principle of the method is described above. However, the difficulty in modeling radiation lies in defining the calculation volume. Indeed, we usually want to determine the far-field radiation, and in this case, the calculation volume is too large. This problem can be circumvented by using a form of the equivalence principle. We define a volume V_C called the *calculation volume*, surrounding the source to a minimum distance which encapsulates the antenna.

The calculation volume is delimited by its surface S. The FEM is applied within the calculation volume, imposing radiation conditions on the surface S. These conditions, in theory, are capable of simulating free space. On the basis of the equivalence principle, when we know the tangential field on the surface, we are able to calculate the field in the whole of the space. This step is known as near-field/far-field transformation. When the far field is calculated, it is easy to determine the antenna's radiation characteristics. The accuracy expected of the method depends on the appropriate choice of the distance from the surface *S* to the antenna.

There are a range of radiation conditions, which are the subject of research in numerical analysis. The principles used are based either on an impedance condition on the surface, or on the representation of the field as a decomposition of modes which are absorbed to a greater or lesser degree, or in the form of a condition which cancels out the waves on the surface, as with the FDTD method with PMLs, to limit the reflections of the waves on *S*.

5.1.3.3.3. Finite Element Method in Time Domain (FEMTD)

The finite element method in time domain is not discussed here. Readers may usefully refer to the *Modern Antenna Handbook*, (Balanis 2008).

5.1.3.3.4. A few strengths of the FEM[12]

– With the FEM, it is relatively simple to process complex geometric forms and non-homogeneous materials.

– It is also simple to handle dispersive materials – i.e. materials whose properties are frequency-dependent.

– The method is capable of handling eigenproblems.

– It is better than the MoM in terms of frequency rise, although the requirement for a mesh model of the whole volume rather than merely the surface means there are more unknowns to deal with.

– It can quite easily be extended to higher-order basis functions: the FEM lends itself well to the use of higher-order basis functions; although complicated, the theoretical extensions are now well understood. It is also possible to use matched elements in order to better approximate curved geometries.

– Multiphysical potential: with the FEM, there is the possibility of coupling EM solutions with, for example, mechanical or thermal solutions. Owing to the widespread popularity and maturity of the FEM in other domains of engineering – that of mechanics, for which the method was originally developed – it is possible to couple computational codes hailing from the domain of electromagnetics, with codes from the domain of mechanics and thermal engineering: for example, detection of hot spots in the case of high powers.

12 (Davidson 2005).

5.1.3.3.5. A few weaknesses of the FEM

– In comparison to the MoM, the processing of highly conductive surfaces is less effective, due to the presence of a mesh between the radiating structure and the absorbent boundary.

– FEM meshes may become extremely complex for large, three-dimensional structures. Indeed, certain works have shown mesh generation times that are equal to or even longer than the time taken to compute the final result.

– The FEM is harder to embed than FDTD. This has a particular impact on its suitability for parallelization, which accounts for the preference for FDTD in "homemade" software.

– Preconditioned iterative solvers are needed when elements of a higher order than the conventional elements are used. This makes it even more complicated to produce a homemade code.

5.1.3.4. *Introduction to the boundary element method (BEM)*[13]

We shall mention this method here because it is frequently used in conjunction with the MoM and FEM.

Boundary Element Method (BEM) codes use the method of moments to solve a problem involving electrical and/or magnetic currents across the surfaces forming the interfaces between two surfaces with the EFIE, MFIE or CFIE equations. Many EMC codes using the MoM also use a boundary element method.

The first step is to represent the geometry of the problem as a distribution of surface currents in a homogeneous medium (usually free space). The fields outside of an object are the incident fields, reflected fields and fields actually radiated by the object. The equivalence theorem states that any source enclosed within a closed volume, radiating an external field across a surrounding surface, allows the source to be replaced with the distribution of equivalent electrical and magnetics fields across that surface.

The problem with the BEM is that it uses surface integrals to solve on all the boundary surfaces. The integrals EFIE and MFIE each include a Green's function, linking the electrical and magnetic surface currents to the electrical and magnetic fields. Green's functions such as these may be complicated to determine for surfaces that, for example, comprise metal surfaces covered in a fine layer of dielectric, or multilayer dielectrics.

---

13 (McDonald 1972; Collin 1991).

In 3D models, the BEM, like the MoM, uses basis functions and test functions which are linear distributions of currents across patches or segments. In general, there are two unknowns per patch, corresponding to two vectorial orthogonal currents. Thin wires may effectively be used with a single unknown representing the amplitude of a current distribution, along each segment of the wire.

In general, BEM techniques excel in modeling unbounded problems, when it is not necessary to model regions with a high degree of complexity.

5.1.3.4.1. A few strengths of the BEM

– Excellent in modeling non-bounded radiation problems

– Excellent for modeling metal surfaces and thin wires

– Good for modeling structures comprising circuit elements

– Effective in CFIE for structures with enclosed surfaces

5.1.3.4.2. A few weaknesses of the BEM

– Not very effective in modeling complex and non-homogeneous materials

– Not highly effective in dealing with problems containing small objects to detail, combined with large objects

– The CFIE formulation is required to model resonating enclosed structures

Above, we placed emphasis on the three numerical methods used in the so-called low-frequency domain, which serve as the foundation for practically all methods developed subsequently in commercial codes, which are therefore particularly reliable. These codes are generally hybridized with various calculation techniques such as the FEBI (Finite Element Boundary Integral). There is the MLFMM (Multilevel Fast Multipole Method), based on the method of moments, which can be combined with the UTD (Uniform Geometrical Theory of Diffraction) to form the referenced method FEBI-MLFMM-UTD, etc. These methods, based on fast solvers, have become more accurate, whilst become increasingly sophisticated, because they are suited to targeted applications, often with much larger dimensions, as shown by Figure 5.2 (FEKO). Among other examples, the FIT (Finite Integration Technique) – a robust method advanced by Weiland (1977) – is a method based on interlocking of the mesh specific to the FDTD, which can be considered a generalization of the FDTD method. It is also similar to the finite element method (FEM). For an example of its performance, in computing structures that are electrically much longer, such as land vehicles or sea vessels (see Figure 5.2), the simulation of a military airplane at the frequency of 960 MHz, with 100,000 unknowns, takes 150 GB of memory with the conventional MoM, compared to 1 GB with the MLFMM (Rokhlin 1993; Chew 2001; Valamparambil 2003).

A helpful overview of numerical methods is given by Sumithra and Thiripurasundari (Sumithra 2017).

5.1.4. *Introduction to high-frequency methods*[14]

The major problem when the size of the antennas increase in relation to the wavelength, or when their medium or nearby environment modifies the currents by means of electromagnetic coupling, is that the number of unknowns in the above-mentioned methods increases rapidly. Computers are no longer able to process these structures, which surpass their capabilities in terms of computation time, memory and accuracy of the results. *High-frequency techniques* consist of an asymptotic assessment of Maxwell's equations. These asymptotic methods are able to process these large geometries with excellent precision, whilst also dealing with the problem posed by computer capacity. Geometrical optics (GO) and physical optics (PO) are the most widely used techniques due to the results obtained in diverse domains such as radar cross-section (RCS), design of reflector antennas or calculation of radiofrequency coverage (application for mobiles, etc.). The main advantage of GO is that the incident and reflective EM waves are calculated on the basis of conservation of the energy flux along a *tube*, known as a *ray*, between the source point and the observation point. GO is very widely used in the area of mobiles to assess RF coverage and characterize transmission channels in an urban environment, by *ray launching* or *ray tracing* (see Volume 1, Chapter 2). GO and PO are not applied unaltered to deal with diffraction phenomena. GO is supplemented by the geometrical theory of diffraction (GTD) (Keller 1956), where the uniform theory of diffraction (UTD) (Kouyoumjian 1974) deals with the directions of non-convergence in Keller's formulas. On the other hand, the physical theory of diffraction (PTD) (Ufimtsev 1962) is applied jointly with the formulation of PO. The PTD improves PO by taking account of non-uniform surface currents and concentrations of the currents near the edges.

In the problem of reflector antennas, PO is the predominant method (Balanis 2008; Harrington 2001). The PO formulation is based on the determination of an equivalent current density, induced on the lighted face of a surface made of a perfect electrical conductor (PEC), locally planar, meaning with a slow variation of

14 (Keller 1956; Ufimtsev 1962; Collin 1969; Kouyoumjian 1974; Silver 1984; Balanis 2008).

curvature in relation to the wavelength, where it is always possible to distinguish a lighted zone from a shaded zone (hidden from the source) (Figure 5.15).

For PO, the formulation is as follows for conductive surfaces: the expression of the current densities $\vec{J}^{PO}$ and $\vec{M}^{PO}$ due to an incident electrical field $\vec{E}_{incident}$ and magnetic field $\vec{H}_{incident}$ produced by an external source visible from a point on the surface and, in view of the boundary conditions on a perfectly conductive surface, have the expression:

$$\vec{J}^{PO} = \hat{n} \wedge \left(\vec{H}_{incident} + \vec{H}_{reflc}\right) \approx 2.\hat{n} \wedge \vec{H}_{incident}$$

$$\vec{M}^{PO} = -\hat{n} \wedge \left(\vec{E}_{incident} + \vec{E}_{refl}\right) = \vec{0}$$

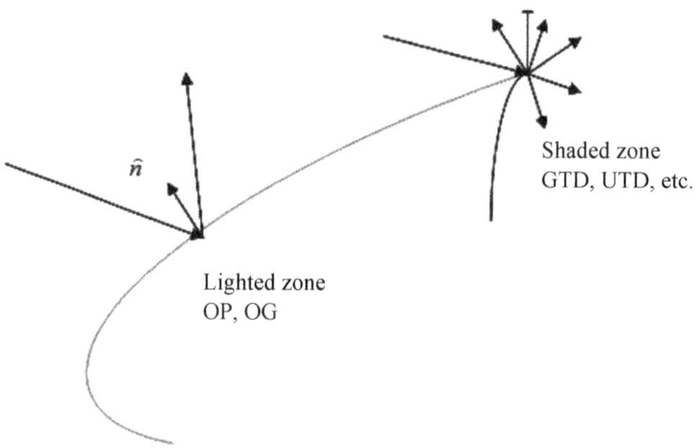

Figure 5.15. *Zone of application of PO and GO on the lighted curved surface of a structure and the zone of application of the GTD or UTD on a segment of an edge*

Consider $\vec{J}^{OP} = 2.\hat{n} \wedge \vec{H}_{incident}$ at a point on the lighted surface; $\vec{J}^{OP} = \vec{0}$ at a point on the hidden surface.

In amplitude and phase, the incident field at a given point on the lighted surface is calculated using the laws of geometrical optics from the source placed at the antenna's phase center.

Physical optics (PO) is based on the currents, unlike geometric optics. It focuses on phenomena where the fundamental nature of light plays a role (particularly its undulating, waveform aspect). PO is based on surface currents, built around the Huygens principle, and takes account of all variations in the wavefront, in both amplitude and phase of the components of the EM field, in any given direction: the effects of diffraction, interference, polarization, aberrations, etc. Thus, it is more widely applicable than GO.

Once the current is known at each point in the structure, the EM field radiated at a point P in the space can be calculated using the equation giving the magnetic field (which is simpler than that given the component of the electrical field):

$$\vec{H}(P) = \frac{1}{\pi} \cdot \iint_s \vec{J}^{OP} \wedge \vec{\nabla} \psi \cdot ds \qquad [5.12]$$

where S is the surface of the lighted zone and $\psi = \frac{e^{-jkr}}{r}$ is the Green's function in the free space.

The electrical field can be deduced in a zone such that $r >> \lambda$ by the plane wave relation:

$$\vec{E}(P) = \sqrt{\frac{\varepsilon_0}{\mu_0}} \hat{R} \wedge \vec{H}(P)$$

where $\hat{R}$ is the unit vector in the direction of the origin of the axes at the observation point P.

The surface integral above, with the exception of simple or symmetrical cases, is never solved analytically. It should be noted that each point of incidence of the field $\vec{H}_{incident}$ giving rise to the current density $\vec{J}^{OP} = 2.\hat{n} \wedge \vec{H}_{incident}$ is totally independent of the other points of incidence, no matter how close they are. Thus, it is possible to sample the conductive surface at any number of surface elements ΔS_i be they triangular, square or of any other shape suited to the geometry of the reflective surface. On these surface elements, we suppose the current to be constant, which enables us to obtain the field radiated for each surface element by a simple Fourier transform:

$$\vec{H}_i(\vec{r}) = \frac{j.k}{4.\pi} \iint_{\Delta S_i} \left(\vec{J}^{OP} \wedge \hat{u} \right) . \frac{e^{-j.k.r}}{r} ds$$

$$= \frac{jk}{4.\pi} \left(\vec{J}^{OP} \wedge \hat{u} \right) . \frac{e^{-j.k.R}}{R} \iint_{\Delta S_i} e^{j.k.(u\alpha + v\beta)} d\alpha . d\beta$$

with $r \approx R-(u.\alpha+v.\beta)$, $u=\sin(\theta).\cos(\varphi)$, $v=\sin(\theta)\sin(\varphi)$, $\hat{u}=\frac{\vec{R}}{R}$ the direction of the observation point from the center of the sample, and α and β the coordinates of a point which describes the surface ΔS_i where the integral operates. The integral is a Fourier transform. In the case that the ΔS_i are rectangles with dimensions a and b, the Fourier transform of a rectangle is the product of two functions in Sin(x)/x, so:

$$\vec{H}_i(\vec{r})=\frac{jk}{4.\pi}\left(\vec{J}^{OP}\wedge\hat{u}\right).\frac{e^{-j.k.R}}{R}..a.b.\frac{\sin(u_1)}{u_1}.\frac{\sin(u_2)}{u_2} \qquad [5.13]$$

with:

$$u_1=\frac{\pi.a}{\lambda}.u \text{ et } u_2=\frac{\pi.b}{\lambda}.v$$

In fact, the samples ΔS_i, or facets, behave like elementary planar apertures.

With the superposition principle, we can calculate the total EM field radiated by the structure by adding together the partial fields $\vec{H}_i(\vec{r})$.

The benefits of asymptotic methods are: the lack of limitation as to the maximum dimension of the structure to be analyzed, the run time which does not increase with rising frequency, and software programs which do not take up too much space in a computer's memory. They take up the baton from *low-frequency* numerical methods when the latter become inoperable. Indeed, the sampling of the reflective surfaces is able to take account of the shifting of those surfaces under their own weight, or due to other factors such as wind, etc.; similarly, the measured or calculated deformations of these surfaces, whatever the cause, can be taken into account.

By way of example, Figure 5.17 shows the radiation pattern of an antenna with offset feed, as represented in Figure 5.16. The sampling of the reflector is a grid measuring $60 \times 60 = 3600$ samples. The computed antenna (Figure 5.17) has a diameter of 1.20 m, its frequency is 10.70 GHz, and the radiation pattern is $\pm 40°$.

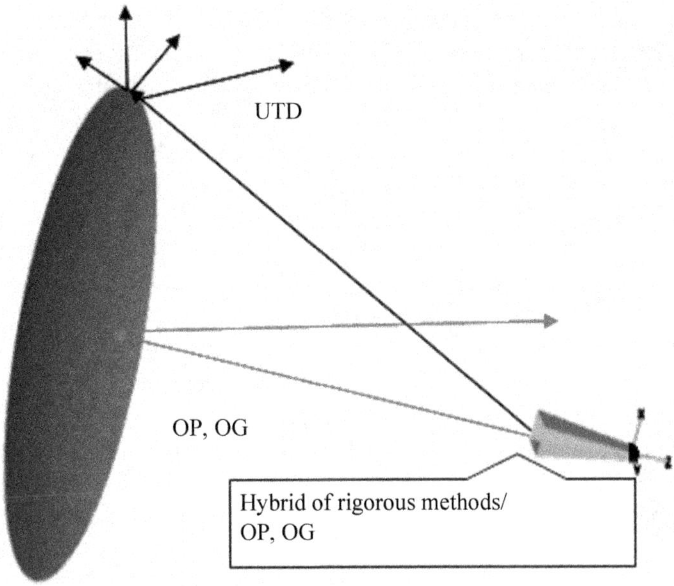

Figure 5.16. *Offset parabolic reflector lit by a horn (Feko 2009)*

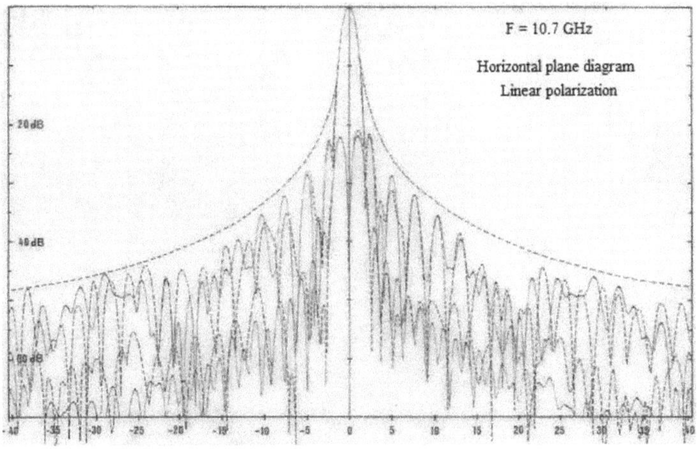

Figure 5.17. *Radiation pattern in nominal polarization and cross-polarization, calculated and measured on a deformed offset antenna 1.20 m in diameter at 10.70 GHz (source: France Télécom)*

Figures 5.18a and 5.18b show the patterns of mode HE11, and tracking of mode TM_{01}, at the output from a periscope with four mirrors (two elliptical and two flat), sent to a Cassegrain antenna 32.5 m in diameter (Pleumeur Bodou (PB4), photo in Figure 2.8b). The periscope is 14.10 m in height and the internal cross-section of the shield is 1.50 m. The number of samples (facets) is 3600 per mirror, giving us: 3600 × 4 = 14,400 in total. The radiation patterns are computed at the shaped subreflector, located 9.90 m away from the center of the planar mirror M4 on the elevation axis, at 30° elevation. The first elliptical mirror M1, in relation to the corrugated horn on the ground, has undergone a 0.20° rotation about its small axis, for Figure 5.18a, and 0.33° for Figure 5.18b.

The directivity of the antenna is calculated for an elevation of 30°, with a total of 21,600 facets for the four mirrors in the periscope and the two reflectors (parabolic and hyperbolic). The directivity is calculated by integration of the radiation pattern (Milligan 2005; Balanis 2008) so D_0= 61.60 dBi, which gives an illumination yield of 80%. The directivity deduced from the measurements of the gain with the INTELSAT procedure (document SSOG 2 dated 10 July 1990) is D_0 = 61.30 dBi.

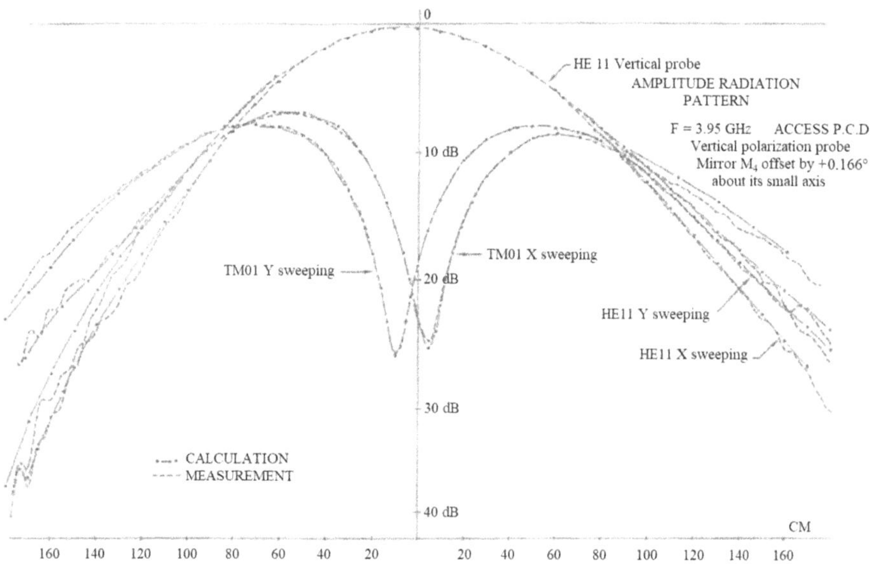

a)

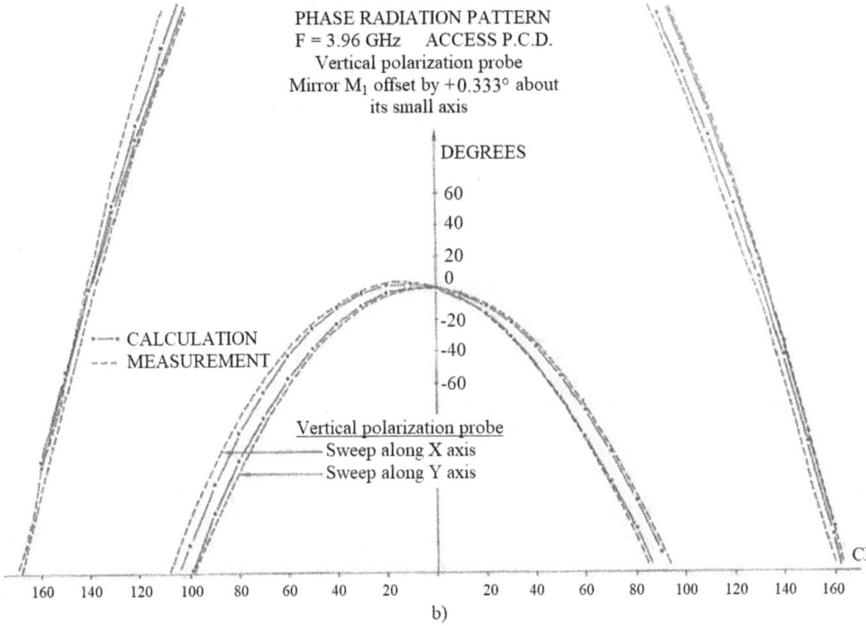

Figure 5.18. *Radiation pattern measured and calculated at the output from a 4-reflector periscope: a) in amplitude of H11 mode and TM01 tracking; b) the phase of HE11 mode (source: France Télécom)*

5.2. General remarks on EMC methods

No EMC method can claim to be able to solve all problems with optimal efficiency. The FEM and FDTD methods are theoretically capable of solving any given problem, but in view of their memory occupation and the prohibitive computation time, it is more suitable to use the method of moments (MoM). However, none of the full-wave (low frequency) methods work in the asymptotic high-frequency domain, given the density of the sampling. Although individual, optimized, non-commercialized methods vary widely in terms of their computer performance, it is clear that in the case of all solvers in the full-wave domain, as the frequency increases, the mesh becomes proportionally finer (and thus the number of unknowns increases).

Also, methods such as adaptive mesh as a function of the working frequency and *fast methods* are a good means of optimizing codes, as is parallelization of supercomputers. Fast methods, including the *Fast Multipole Method* and the

Adaptative Integral Method, are designed to reduce the computing resources and time consumed by the codes.

In summary, these methods replace the traditional algorithms of direct matrix solution by iterative solvers, and use methods to approximate the interaction between the parts of the mesh which are spaced a reasonable distance apart (generally at least several times the wavelength). The matrix-vector product, which is at the heart of iterative solvers, using a fast technique similar to the FFT, reduces the cost from O (N**2) to O (N log N) by iteration (Chew 1997, 2001).

When full-wave techniques with supercomputers ultimately make it impossible to run the calculations for large structures, asymptotic techniques such as PO and GO become important. As we have seen, these methods generally use rays as field propagators and essentially localize the electromagnetic interaction by calculating the field at a point as the sum of the direct, reflected and diffracted rays, originating at all points (or sometimes lines) in the structure. With these methods, there is no concept of discretization of an unknown field, although the surface is divided into patches, on each of which the current is known by the incident field given in GO, rather than based on a matrix equation applying a boundary condition.

Appendix 1

Mathematical Formulae[1]

A1.1. Trigonometric transformation equations

If x and y are any two angles, then the following relationships are verified:

$$\begin{cases} \sin(x \pm y) = \sin(x)\cos(y) \pm \cos(x)\sin(y) \\ \cos(x \pm y) = \cos(x)\cos(y) \mp \sin(x)\sin(y) \end{cases}$$

$$\begin{cases} \sin(x) \pm \sin(y) = 2\sin\dfrac{x \pm y}{2}\cos\dfrac{x \mp y}{2} \\ \cos(x) + \cos(y) = 2\cos\dfrac{x + y}{2}\cos\dfrac{x - y}{2} \\ \cos(x) - \cos(y) = -2\sin\dfrac{x + y}{2}\sin\dfrac{x - y}{2} \end{cases} \qquad [A1.1]$$

$$\begin{cases} \sin 2x = 2\sin x \cos x \\ \cos 2x = \cos^2 x - \sin^2 x \\ \tan 2x = 2\sin\dfrac{2\tan x}{1 - \tan^2 x} \end{cases} ; \begin{cases} 2\sin^2\dfrac{x}{2} = 1 - \cos x \\ 2\cos^2\dfrac{x}{2} = 1 + \cos x \\ \cos^2(x) + \sin^2(x) = 1 \\ \cos^2(x) = \dfrac{1}{1 + \tan^2 x} \end{cases} \qquad [A1.2]$$

For a color version of all figures in this book, see www.iste.co.uk/favennec/electromagnetic2.zip.

1 Bok, J., Hulin-Jung, N. (1979). *Relativité, ondes électromagnétiques.* Hermann, Paris; Bruneaux, J., Saint-Jean, M., Matricon, J. (2002). *Électrostatique et magnétostatique*. Belin, Paris; Amzallag, E., Cipriani, J., Aïm, J.B., Piccioli, N. (2006). *Électrostatique et électrocinétique.* Dunod, Paris.

Electromagnetic Waves 2,
coordinated by Pierre-Noël FAVENNEC. © ISTE Ltd 2020

Moreover, the hyperbolic functions (*hyperbolic sine sinh, hyperbolic cosine cosh and hyperbolic tangent tanh functions*) are given by the following expressions:

$$\sinh x = \frac{e^x - e^{-x}}{2}\,; \ \cosh x = \frac{e^x + e^{-x}}{2}\,; \ \tanh x = \frac{e^x - e^{-x}}{e^x + e^{-x}} \qquad [A1.3]$$

The imaginary number i verifies property $i^2 = -1$. For a given angle x, we obtain:

$$e^{ix} = \cos x + i \sin x\,; \ \sin x = \frac{e^{ix} - e^{-ix}}{2i}\,; \ \cos x = \frac{e^{ix} + e^{-ix}}{2} \qquad [A1.4]$$

A1.2. Series developments

If x is a variable, the following series developments are satisfied:

– *binomial development*:

$$(1+x)^n = 1 + nx + \frac{n(n-1)}{2!}x^2 + \frac{n(n-1)(n-2)}{3!}x^3 + \ldots\ldots \qquad [A1.5]$$

$$\text{if } x << 1\text{: } (1+x)^n \approx 1 + nx \qquad [A1.6]$$

– *Maclaurin series* (Colin Maclaurin (1698–1746), Scottish mathematician):

$$e^x = 1 + x + \frac{x^2}{2!} + \frac{x^3}{3!} + \frac{x^4}{4!} + \ldots\ldots;\ \text{if } x << 1\text{: } e^x \approx 1 + x \qquad [A1.7]$$

$$a^x = e^{x \ln a} = 1 + \frac{(\ln a)}{1!}x + \frac{(\ln a)^2}{2!}x^2 + \frac{(\ln a)^3}{3!}x^3 + \ldots\ldots \qquad [A1.8]$$

$$\text{if } x << 1\text{: } a^x = 1 + \frac{(\ln a)}{1!}x \qquad [A1.9]$$

$$\ln(1+x) = x - \frac{x^2}{2} + \frac{x^3}{3} - \frac{x^4}{4} + \ldots\ldots \qquad [A1.10]$$

$$\text{if } x << 1\text{: } \ln(1+x) \approx x \qquad [A1.11]$$

$$\sin x = x - \frac{x^3}{3!} + \frac{x^5}{5!} - \frac{x^7}{7!} + \ldots\ldots; \text{ if } x << 1: \ \sin x \approx x \qquad [A1.12]$$

$$\cos x = 1 - \frac{x^2}{2!} + \frac{x^4}{4!} - \frac{x^6}{6!} + \ldots\ldots; \text{ if } x << 1: \ \cos x \approx 1 \qquad [A1.13]$$

$$\tan x = x + \frac{x^3}{3} + \frac{2x^5}{15} + \frac{17x^7}{315} + \ldots\ldots; \text{ if } x << 1: \ \tan x \approx x \qquad [A1.14]$$

– *Taylor series* (Brook Taylor (1685–1731), English mathematician): for a given function $f(x)$, the Taylor series development about x_0 is written:

$$f(x) = f(x_0) + (x - x_0)\left(\frac{df}{dx}\right)_{x=x_0} + \frac{(x-x_0)^2}{2!}\left(\frac{d^2 f}{dx^2}\right)_{x=x_0} + \ldots + \frac{(x-x_0)^n}{n!}\left(\frac{d^n f}{dx^n}\right)_{x=x_0} \qquad [A1.15]$$

Appendix 2

Vector Calculations

A2.1. Vectors in coordinate systems

A2.1.1. *Cartesian coordinate systems*

In 3D Cartesian space, a point M is described by its Cartesian coordinates x, y and z in a direct orthonormal trihedron of basic vectors generally denoted $\vec{e}_x$, $\vec{e}_y$ and $\vec{e}_z$ (Figure A2.1).

In the Cartesian coordinate system:

– the elementary volume is: $d\tau = dxdydz$;

– the position vector $\overrightarrow{OM}$ (Figure A2.1) is defined by the relation:

$$\overrightarrow{OM} = x\vec{e}_x + y\vec{e}_y + z\vec{e}_z \qquad [A2.1]$$

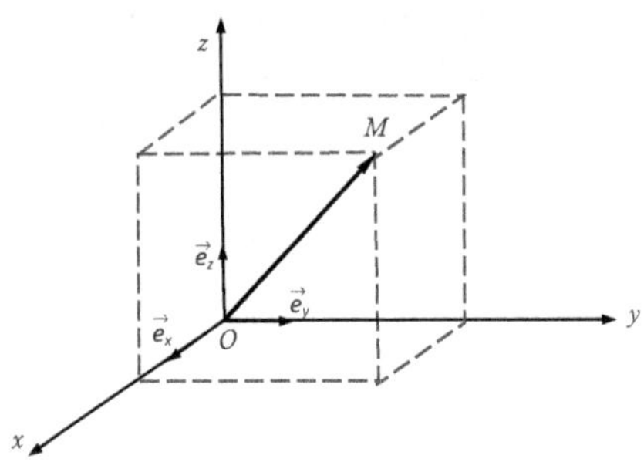

Figure A2.1. *Cartesian coordinate system*

For a color version of all figures in this book, see www.iste.co.uk/favennec/electromagnetic2.zip.

Electromagnetic Waves 2,
coordinated by Pierre-Noël FAVENNEC.

Any vector $\vec{G}$ of components G_x, G_y and G_z is written in Cartesian space dimensions:

$$\vec{G} = G_x \vec{e}_x + G_y \vec{e}_y + G_z \vec{e}_z \quad \text{[A2.2]}$$

Any displacement vector $\vec{dl} = d\overrightarrow{MM'}$ between two very close points M and M' is written:

$$\vec{dl} = dx\vec{e}_x + dy\vec{e}_y + dz\vec{e}_z \quad \text{[A2.3]}$$

A2.1.2. *Cylindrical coordinate systems*

If the studied problem presents a cylindrical symmetry compared to an axis of revolution (Oz example), the cylindrical coordinate system is more adapted to the studied problem (case of a rectilinear wire traversed by an electric current where one wants to determine the magnetic field which it creates in a point of space, the point being located on a horizontal plane perpendicular to the wire arranged vertically). In this system, the position of point M is identified by three coordinates (Figure 1.37):

– its distance ρ from the origin O;

– its side z with respect to the xOy plane;

– the azimuth angle θ between the xz plane and the vertical plane containing point M and axis Oz; θ varies between 0 and π.

In the cylindrical coordinate system, a point M (ρ, θ, z) is located in a direct orthonormal local trihedron of basis vectors denoted $\vec{e}_\rho$, $\vec{e}_\theta$ and $\vec{e}_z$. The rotation vectors are such that $\vec{e}_\rho$ is perpendicular to the Oz axis and $\vec{e}_\theta$ is perpendicular to the vertical plane containing point M and the Oz axis. They are oriented as shown in Figure A2.2.

In the cylindrical coordinate system:

– the elementary volume is: $d\tau = \rho d\rho d\theta dz$;

– the position vector $\overrightarrow{OM}$ is given by the relation:

$$\overrightarrow{OM} = \rho\vec{e}_\rho + z\vec{e}_z \quad \text{[A2.4]}$$

– in the orthonormal basis ($\vec{e}_\rho$, $\vec{e}_\theta$, $\vec{e}_z$), any vector $\vec{G}$ of components (G_ρ, G_θ, G_z) is written:

$$\vec{G} = G_\rho \vec{e}_\rho + G_\theta \vec{e}_\theta + G_z \vec{e}_z \qquad \text{[A2.5]}$$

– a displacement vector $d\vec{l} = d\overrightarrow{MM'}$ between two very close points M and M' is written:

$$d\vec{l} = d\rho \vec{e}_\rho + \rho d\theta \vec{e}_\theta + dz \vec{e}_z \qquad \text{[A2.6]}$$

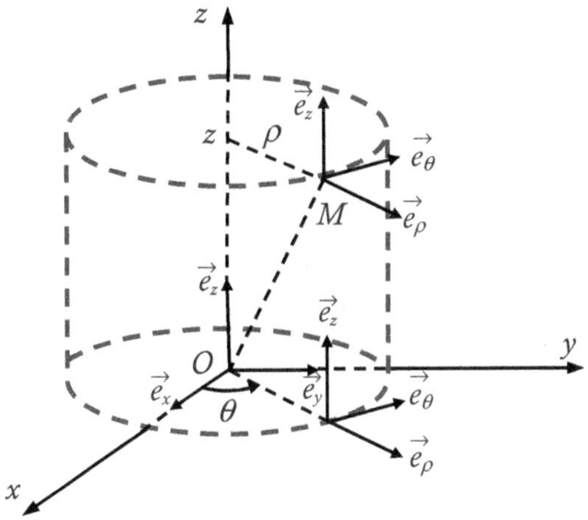

Figure A2.2. *Cylindrical coordinate system*

To understand the origin of the term $\rho d\theta \vec{e}_\theta$ in [A2.6], we derive [A2.4], with respect to time:

$$\frac{d\vec{l}}{dt} = \frac{d\overrightarrow{OM}}{dt} = \frac{d(\rho \vec{e}_\rho)}{dt} + \frac{d(z\vec{e}_z)}{dt} = \frac{d\rho}{dt}\vec{e}_\rho + \frac{\rho d(\vec{e}_\rho)}{dt} + \frac{dz}{dt}\vec{e}_z$$

i.e.:

$$\frac{d\vec{l}}{dt} = \frac{d\rho}{dt}\vec{e}_\rho + \rho \frac{d(\vec{e}_\rho)}{d\theta}\frac{d\theta}{dt} + \frac{dz}{dt}\vec{e}_z = \frac{d\rho}{dt}\vec{e}_\rho + \rho \vec{e}_\theta \frac{d\theta}{dt} + \frac{dz}{dt}\vec{e}_z \qquad \text{[A2.7]}$$

By removing dt in the last equality in [A2.7] we find [A2.6].

Note that we used the definition of the derivative of a rotation vector: the derivative of a rotation vector with respect to the angle of rotation is equal to the vector which is directly perpendicular to it by going in the direction of rotation. So considering Figure A2.2, we obtain:

$$\frac{d(\vec{e}_\rho)}{d\theta} = \vec{e}_\theta ; \; \frac{d(\vec{e}_\theta)}{d\theta} = -\vec{e}_\rho \qquad \text{[A2.8]}$$

A2.1.3. *Spherical coordinate systems*

If the problem studied has a spherical symmetry with respect to the origin of the coordinates, the spherical coordinate system is more appropriate for the problem studied. In this system, the position of point M is identified by three coordinates:

– its distance r from the origin O;

– the azimuth angle φ (or azimuth) between the xz plane and the vertical plane containing point M and the Oz axis; φ varies between 0 and 2π;

– the angle θ (colatitude or zenith) between the $\overrightarrow{OM}$ radius and the Oz *axis*; θ varies between 0 and π.

In the spherical coordinate system, any point M (r, θ, φ) is located in a direct orthonormal local trihedron of basis vectors denoted $\vec{e}_r$, $\vec{e}_\theta$ and $\vec{e}_\varphi$. These three rotation vectors are oriented as shown in Figure A2.3.

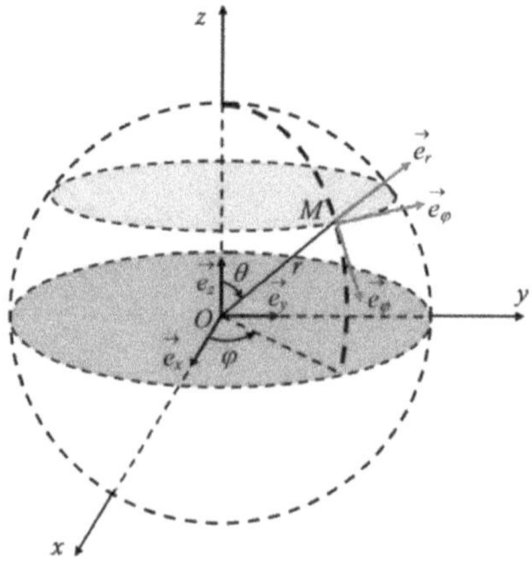

Figure A2.3. *Spherical coordinate systems*

In the spherical coordinate system:

– the elementary volume is: $d\tau = r^2 dr \sin\theta d\theta d\varphi$;

– the position vector $\overrightarrow{OM}$ is given by the relation:

$$\overrightarrow{OM} = r\vec{e}_r \quad [A2.9]$$

– in the orthonormal basis ($\vec{e}_r$, $\vec{e}_\theta$, $\vec{e}_\varphi$), any vector $\vec{G}$ of components G_r, G_θ and G_φ is written:

$$\vec{G} = G_\rho \vec{e}_\rho + G_\theta \vec{e}_\theta + G_\varphi \vec{e}_\varphi \quad [A2.10]$$

– any displacement vector $\vec{dl} = \overrightarrow{dMM'}$ between two very close points M and M' is written:

$$d\vec{l} = dr\vec{e}_r + rd\theta\vec{e}_\theta + r\sin\theta d\varphi\vec{e}_\varphi \quad [A2.11]$$

Cartesian coordinates (x, y, z) are linked to spherical coordinates (r,θ, φ) by the following relations. These relations can be deduced from Figure A2.3:

$$\begin{cases} x = r\sin\theta\cos\varphi \\ y = r\sin\theta\sin\varphi\ ;\ OM = r = \sqrt{x^2 + y^2 + z^2} \\ z = r\cos\theta \end{cases} \quad [A2.12]$$

A2.1.4. *Laws of orientation in space*

A2.1.4.1. *Notion of direct trihedron*

Let us consider a trihedron formed by the Cartesian basis vectors $\vec{e}_x$, $\vec{e}_y$ and $\vec{e}_z$ (Figure A2.4). By definition, the trihedron ($\vec{e}_x$, $\vec{e}_y$, $\vec{e}_z$) is direct if, for vector $\vec{e}_z$, the rotation of the vector $\vec{e}_x$ on vector $\vec{e}_y$ according to the angle less than 180° occurs in the trigonometric sense.

The trihedron formed from three fingers on the right hand (thumb, index, middle finger) is very practical for determining the direction of Laplace's magnetic force or that of Lorentz's magnetic force.

A2.1.4.2. *Orientation of the surface vector*

For a surface S based on any contour (C), we define a surface vector $\overrightarrow{dS}$ whose modulus is equal to the area dS of the elementary surface, of direction normal to the surface and of direction directly associated with the direction of travel indicated by the corkscrew rule.

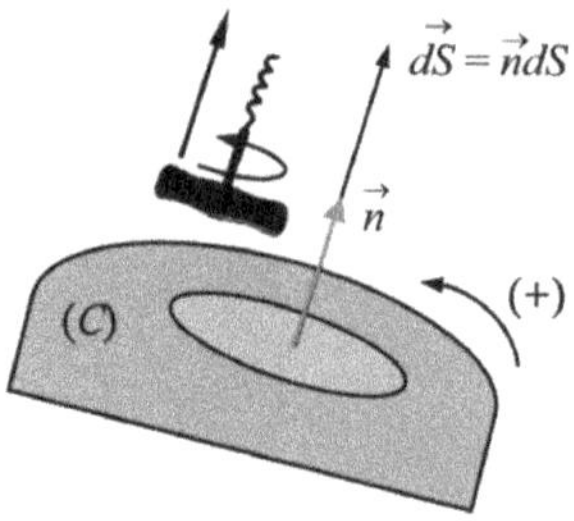

Figure A2.4. *Law of orientation of an elementary surface, corkscrew rule*

If $\vec{n}$ is the vector normal to the surface, then $\overrightarrow{dS} = \vec{n}\,dS$. By arbitrarily choosing a positive direction of travel on the contour (C), the direction of the surface vector is the direction in which the corkscrew advances when it is turned in the chosen positive direction (Figure A2.4).

A2.1.5. *Solid angle*

An angle (in radians) is defined as the ratio of the length l of the arc intersected by the circle, centered at the apex angle, and radius R of this circle, i.e. $\theta = l/R$.

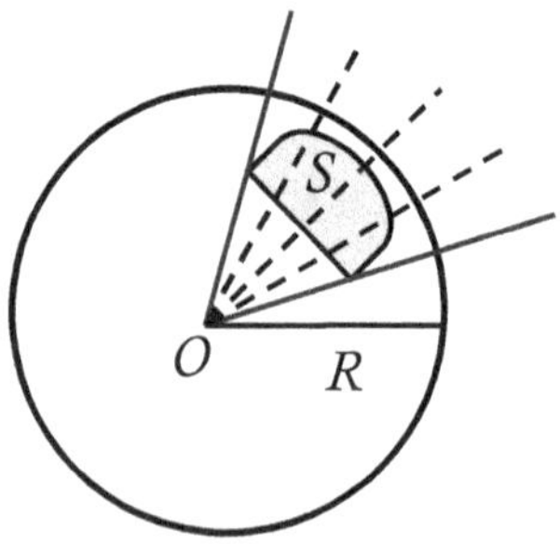

Figure A2.5. *Surface* S *intersected by a cone on a sphere with radius* R

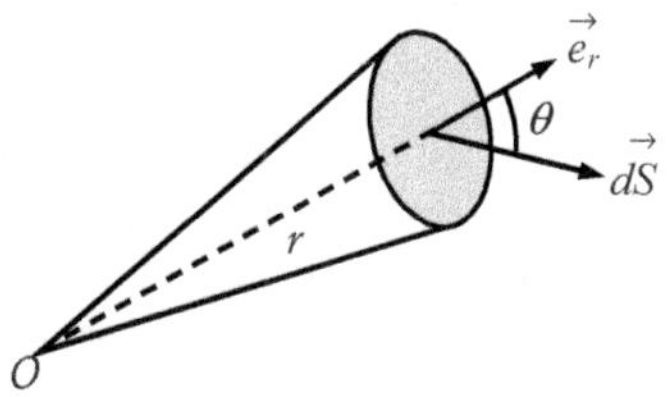

Figure A2.6. *Surface* dS *seen from point* O *at a distance* r

By analogy, we define the solid angle Ω (in *steradians*) of a cone as the ratio of the surface S intersected by this cone on a sphere centered at its apex and the square of the radius of the sphere (Figure A2.5):

$$\Omega = \frac{S}{R^2} \qquad [A2.13]$$

– for any space, $S = \pi R^2 \Rightarrow \Omega = 4\pi$ steradians;

– for a demi-space, $S = 2\pi R^2 \Rightarrow \Omega = 2\pi$ steradians.

Very often, we are led to take an interest in the solid angle which we see from O, an elementary surface dS of small dimensions compared to the distance r which separates the surface of O, and whose normal makes an angle θ with the direction of the unit vector $\vec{e_r}$ (Figure A2.6). This elementary solid angle intervenes for example in the definition of the flow of a vector field through an elementary surface dS. By definition, the elementary solid angle is given by the following relationship:

$$d\Omega = \frac{\vec{dS} \cdot \vec{e_r}}{r^2} = \frac{dS \cos\theta}{r^2} \qquad [A2.14]$$

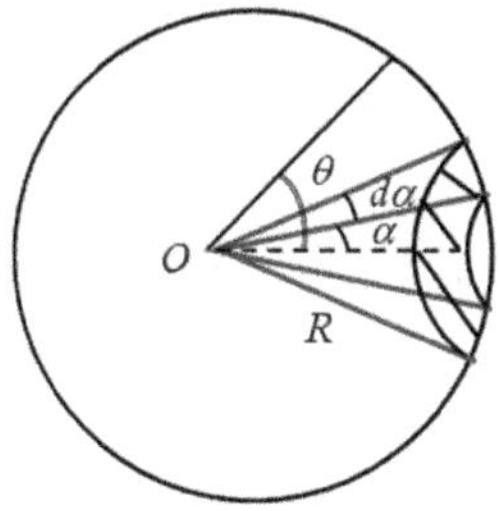

Figure A2.7. *Cones with demi-angles at the apex* α *and* $\alpha + d\alpha$

We express the solid angle Ω of a cone of revolution with apex O and demi-angle at the apex θ. For this, we consider two cones of revolution, with the same apex O, the same axis, and demi-angles α and $\alpha + d\alpha$ (Figure A2.7).

The surface (hatched) cut on the sphere with radius R by the interval between the two cones is given by the relation:

$$dS = 2\pi R \sin\alpha \, R d\alpha = 2\pi R^2 \sin\alpha d\alpha$$

The corresponding elementary solid angle $d\Omega$ is therefore:

$$d\Omega = \frac{dS}{R^2} = 2\pi \sin\alpha d\alpha \qquad [A2.15]$$

Integrating [A2.15] between the limits 0 and θ, we obtain the solid angle Ω:

$$\Omega = 2\pi \int_0^{\theta} \sin\alpha d\alpha = 2\pi(1-\cos\theta) \qquad [A2.16]$$

Similarly, we find:

– for any space, $\theta = \pi \Rightarrow \Omega = 4\pi$ steradians;

– for a demi-space, $\theta = \pi/2 \Rightarrow \Omega = 2\pi$ steradians.

A2.1.6. *Scalar product of two vectors*

Let us consider two vectors $\vec{G_1}$ and $\vec{G_2}$ forming between them an angle θ (Figure A2.8).

By definition, the scalar product of these vectors is the scalar P such that:

$$P = \overrightarrow{G_1} \cdot \overrightarrow{G_2} = G_1.G_2 \cos\theta \qquad [A2.17]$$

The scalar product is equal to zero if $\vec{G_1}$ and $\vec{G_2}$ are orthogonal. This is the case of *basis vectors of direct trihedra* of Cartesian, cylindrical and spherical coordinate systems.

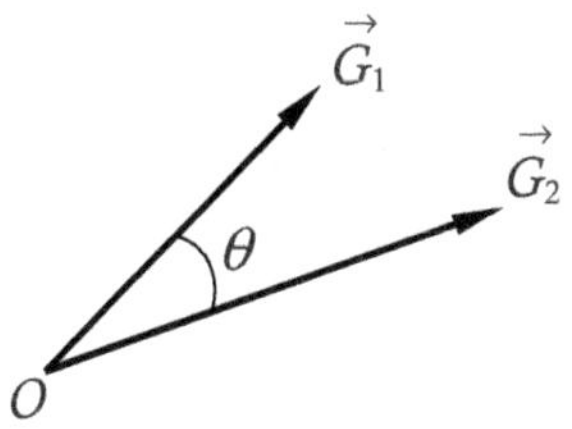

Figure A2.8. *Vectors $\vec{G}_1$ and $\vec{G}_2$ forming angle θ between them*

A2.1.7. *Vector product of two vectors*

Let us consider vectors $\vec{G}_1$ and $\vec{G}_2$ (Figure A2.9). By definition, the vector product of these vectors is the vector $\vec{P}$ such that:

$$\vec{P} = \vec{G_1} \wedge \vec{G_2} \quad \text{[A2.18]}$$

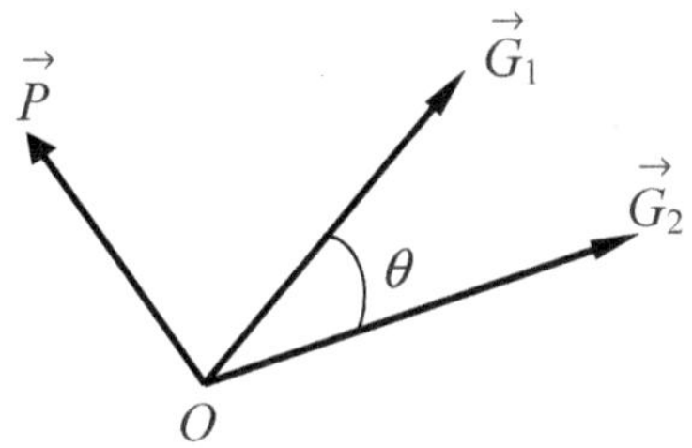

Figure A2.9. *Vector product* P *of two vectors $\vec{G}_1$ and $\vec{G}_2$*

The vector product $\vec{P}$ is perpendicular to the plane formed by the two vectors $\vec{G}_1$ and $\vec{G}_2$, and its sense is such that the trihedron ($\vec{G}_1$, $\vec{G}_2$, $\vec{P}$) is direct (Figure A2.9) and its norm is equal to $G_1G_2|\sin\theta|$. The vector product $\vec{P}$ is null if $\theta = 0$ or π. Moreover, note that the vector product is very important in electromagnetism. It appears in expressions of Laplace's law and the Lorentz law.

A2.1.8. *Field*

A field is the association of a value (of a tensor in the broad sense) of a physical magnitude at each point in space. The space considered may or may not be occupied by a material object.

We distinguish:

– scalar fields (order 0 tensor): at each point in space, we associate a number, for example temperature, pressure, density, etc.;

– vector fields (order 1 tensor): at each point in space we associate a vector, for example the gravity field vector, the electric field vector, the local velocity of a fluid etc.;

– matrix fields (order 2 tensor): at each point, we associate a matrix, for example the elastic deformation tensor, strain tensors.

This notion of a field is most often adapted to the study of continuous environments (mechanics of continuous environments, fluid mechanics) as well as to electromagnetic phenomena. This is essential for the efficient processing of wave phenomena.

We call a field line any "oriented curve" of space which is the tangent at any point to the associated field and oriented in the same direction as the latter. Drawing the field lines establishes the general appearance of the field in a given region of space.

These different fields are represented by functions. When these functions have partial derivatives, we can define other scalars or vectors which remain the same whatever the frame of reference: they are invariants of the scalar field or of the vector field. They provide intrinsic and local properties.

For a *scalar field*, the invariants are the gradient (a vector) and the Laplacian (a scalar).

For a *vector field*, the invariants are the rotational (a vector), the divergence (a scalar) and the vector Laplacian (a vector).

A2.1.9. *Circulation of a vector*

Let us consider that $\vec{G}$ is a vector field and (C) is a circuit whose path is orientated (Figure A2.10).

By definition, the circulation C of the vector field on the circuit is given by:

$$dC = \vec{G} \cdot \vec{dl} \Rightarrow C = \int_{(C)} \vec{G} \cdot \vec{dl} \qquad \text{[A2.19]}$$

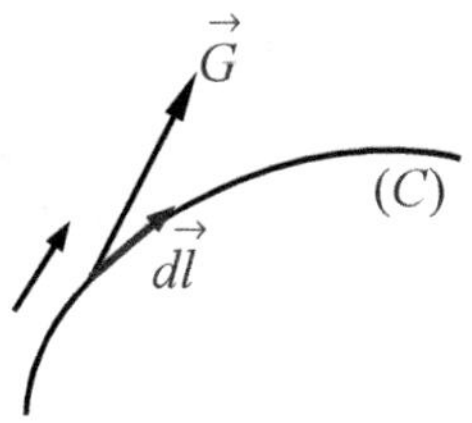

Figure A2.10. *Circulation of a vector on a circuit (C)*

If the considered circuit is closed, then the circulation [A2.19] is written:

$$C = \oint_{(C)} \vec{G} \cdot \vec{dl} \qquad \text{[A2.20]}$$

The small circle on the symbol of the integral shows that the integration occurs along the whole circuit.

NOTE.– In the case where the vector field is a force field $\vec{F}$, then the expression [A2.19] gives:

$$C = \int_{(C)} \vec{F} \cdot \vec{dl} = W(\vec{F}) \qquad \text{[A2.21]}$$

The result [A2.21] shows that the circulation of a force field on a circuit is nothing other than the work done by the force field on the circuit (for example the work of the weight of a body).

A2.1.10. *Flux of a vector*

Let us consider any vector field $\vec{G}$ and $\vec{dS}$ an elementary surface. The elementary flux of the field $\vec{G}$ leaving the elementary surface is given by the expression:

$$d\Phi = \vec{G} \cdot \vec{dS} \qquad \text{[A2.22]}$$

If $\vec{dS} = \vec{n}dS$ is based on a contour (C), the total flux leaving the surface S is:

$$\Phi = \iint_S \vec{G} \cdot \vec{dS} = \iint_S \vec{G} \cdot \vec{n}dS \qquad \text{[A2.23]}$$

The conventions on the orientation of the surface vector were prevoiusly specified (see section A2.1.4.2).

A2.2. Vector operators

A2.2.1. *Gradient operators*

The nabla $\vec{\nabla}$ vector operator is the gradient operator associating a scalar function f and a vector $\overrightarrow{grad}\, f$ (or $\vec{\nabla} f$) such that for an elementary displacement $\overrightarrow{dl}$:

$$df = \vec{\nabla} f \cdot \overrightarrow{dl} \qquad [A2.24]$$

The gradient is an operator which takes an input scalar field and returns a vector. It defines the slope (direction of the greatest variation) of this scalar field at this point.

The nabla operator is involved in many vector calculations in the particular field of electromagnetism which interests us in this book. The expressions for the gradient operator in the various common coordinate systems are as follows.

A2.2.1.1. *Cartesian coordinates*

Scalar function $f(x, y, z)$; direct trihedron ($\vec{e}_x$, $\vec{e}_y$, $\vec{e}_z$):

$$\vec{\nabla} f = \frac{\partial f}{\partial x}\vec{e}_x + \frac{\partial f}{\partial y}\vec{e}_y + \frac{\partial f}{\partial z}\vec{e}_z \qquad [A2.25]$$

A2.2.1.2. *Cylindrical coordinates*

Scalar function $f(\rho, \theta, z)$; direct trihedron ($\vec{e}_\rho$, $\vec{e}_\theta$, $\vec{e}_z$):

$$\vec{\nabla} f = \frac{\partial f}{\partial \rho}\vec{e}_\rho + \frac{1}{\rho}\frac{\partial f}{\partial \theta}\vec{e}_\theta + \frac{\partial f}{\partial z}\vec{e}_z \qquad [A2.26]$$

A2.2.1.3. *Spherical coordinates*

Scalar function $f(r, \theta, \varphi)$; direct trihedron ($\vec{e}_r$, $\vec{e}_\theta$, $\vec{e}_\varphi$):

$$\vec{\nabla} f = \frac{\partial f}{\partial r}\vec{e}_r + \frac{1}{r}\frac{\partial f}{\partial \theta}\vec{e}_\theta + \frac{1}{r\sin\theta}\frac{\partial f}{\partial \varphi}\vec{e}_\varphi \qquad [A2.27]$$

For a path AB along a circuit (C), the circulation of the gradient of a scalar function f is given by the expression:

$$C = \int_{AB} \vec{\nabla} f . \vec{dl} = \int_{A}^{B} df = f(A) - f(B) \qquad \text{[A2.28]}$$

The result [A2.28] shows that the gradient of the scalar function f does not depend on the path followed. In the case of a closed circuit, the circulation of the gradient of a scalar function f is therefore zero:

$$C = f(A) - f(A) = 0$$

A2.2.2. *Divergence operator*

The divergence vector operator div (or even $\vec{\nabla}\cdot$) associates a scalar, denoted div. $\vec{G}$ (or even $\vec{\nabla}\cdot\vec{G}$), to a vector field $\vec{G}$. For a Cartesian coordinate system, the divergence of field $\vec{G}$ is equal to the scalar product of field $\vec{G}$ and the operator nabla $\vec{\nabla}$, i.e.:

$$\text{div}\, \vec{G} = \vec{\nabla}\cdot\vec{G} \qquad \text{[A2.29]}$$

The divergence is an operator that takes an input vector field and returns a scalar.

It represents the flux (per unit volume) of this field (across the surface delimiting a unit of volume at this point).

A *positive divergence* at point M (x,y,z) corresponds to a mainly outgoing flow around this point (divergent field: expansion of a fluid for example).

A *negative divergence* at point M (x,y,z) corresponds to a mainly incoming flow around this point (convergent field: fluid compression for example).

A *zero divergence* at point M (x,y,z) corresponds to incoming and outgoing flows around this point which compensate each other (uniform field: incompressible fluid) or to a swirling flow field.

Like the nabla operator, the divergence operator is involved in many vector calculations in the field of electromagnetism (for example the Green-Ostrogradsky theorem). The expressions of the divergence operator in the case of the various coordinate systems are as follows.

A2.2.2.1. *Cartesian coordinates*

Vector field $\vec{G}$; direct trihedron ($\vec{e}_x$, $\vec{e}_y$, $\vec{e}_z$):

$$\vec{\nabla}\cdot\vec{G}=\frac{\partial G_x}{\partial x}+\frac{\partial G_y}{\partial y}+\frac{\partial G_z}{\partial z} \qquad [A2.30]$$

A2.2.2.2. *Cylindrical coordinates*

Vector field $\vec{G}$; direct trihedron ($\vec{e}_\rho$, $\vec{e}_\theta$, $\vec{e}_z$):

$$\vec{\nabla}\cdot\vec{G}=\frac{1}{\rho}\frac{\partial(\rho G_\rho)}{\partial\rho}+\frac{1}{\rho}\frac{\partial G_\theta}{\partial\theta}+\frac{\partial G_z}{\partial z} \qquad [A2.31]$$

A2.2.2.3. *Spherical coordinates*

Vector field $\vec{G}$; direct trihedron ($\vec{e}_r$, $\vec{e}_\theta$, $\vec{e}_\varphi$):

$$\vec{\nabla}\cdot\vec{G}=\frac{1}{r^2}\frac{\partial(r^2G_r)}{\partial r}+\frac{1}{r\sin\theta}\frac{\partial(\sin\theta\, G_\theta)}{\partial\theta}+\frac{1}{r\sin\theta}\frac{\partial G_\varphi}{\partial\varphi} \qquad [A2.32]$$

A2.2.3. *Rotation operator*

The rotation vector operator $\overrightarrow{rot}$ (or even $\vec{\nabla}\wedge$) associates vector $\vec{\nabla}\wedge\vec{G}$ to a field of vectors $\vec{G}$. For a Cartesian coordinate system, the rotational of a vector field $\vec{G}$ is equal to the vector product of the field $\vec{G}$ and the nabla operator $\vec{\nabla}$, i.e.:

$$\overrightarrow{rot}\ \vec{G}=\vec{\nabla}\wedge\vec{G} \qquad [A2.33]$$

The rotational is a partial derivative operator which associates another vector field to a three-dimensional vector field $G\,(x,y,z)$.

It expresses the tendency of the field lines of a vector field to rotate around a point. The presence of this vector means that this field rotates around this axis. The amplitude of the vector which characterizes this rotation will be greater, the greater the rotation.

Like the nabla and divergence operators, the rotational operator is involved in many vector calculations in the field of electromagnetism (for example, Stokes' theorem). The expressions of the rotational operator in the various coordinate systems are as follows.

A2.2.3.1. *Cartesian coordinates*

Vector field $\vec{G}$ (G_x,G_y,G_z); direct trihedron ($\vec{e}_x$, $\vec{e}_y$, $\vec{e}_z$):

$$\vec{\nabla} \wedge \vec{G} = \left(\frac{\partial G_z}{\partial y} - \frac{\partial G_y}{\partial z} \right) \vec{e}_x + \left(\frac{\partial G_x}{\partial z} - \frac{\partial G_z}{\partial x} \right) \vec{e}_y + \left(\frac{\partial G_y}{\partial x} - \frac{\partial G_x}{\partial y} \right) \vec{e}_z \qquad \text{[A2.34]}$$

A2.2.3.2. *Cylindrical coordinates*

Vector field $\vec{G}$ (G_ρ,G_θ,G_z); direct trihedron ($\vec{e}_\rho$, $\vec{e}_\theta$, $\vec{e}_z$):

$$\vec{\nabla} \wedge \vec{G} = \left(\frac{1}{\rho} \frac{\partial G_z}{\partial \theta} - \frac{\partial G_\theta}{\partial z} \right) \vec{e}_\rho + \left(\frac{\partial G_\rho}{\partial z} - \frac{\partial G_z}{\partial \rho} \right) \vec{e}_\theta + \left(\frac{1}{\rho} \frac{\partial (\rho G_\theta)}{\partial \rho} - \frac{\partial G_\rho}{\partial \theta} \right) \vec{e}_z \qquad \text{[A2.35]}$$

A2.2.3.3. *Spherical coordinates*

Vector field $\vec{G}$ (G_r,G_θ,G_φ); direct trihedron ($\vec{e}_r$, $\vec{e}_\theta$, $\vec{e}_\varphi$):

$$\vec{\nabla} \wedge \vec{G} = \frac{1}{r \sin\theta} \left(\frac{\partial (\sin\theta G_\varphi)}{\partial \theta} - \frac{\partial G_\theta}{\partial \varphi} \right) \vec{e}_r + \left(\frac{1}{r \sin\theta} \frac{\partial G_r}{\partial \varphi} - \frac{1}{r} \frac{\partial (r G_\varphi)}{\partial r} \right) \vec{e}_\theta + \left(\frac{1}{r} \frac{\partial (r G_\theta)}{\partial r} - \frac{1}{r} \frac{\partial G_r}{\partial \theta} \right) \vec{e}_\varphi \qquad \text{[A2.36]}$$

A2.2.4. *Laplacian operator*

The Laplacian operator, or simply the Laplacian, is a vector operator denoted ∇^2 (or even $\vec{\nabla} \cdot \vec{\nabla} = \Delta$). It associates a scalar f to another scalar, denoted $\vec{\nabla} \cdot (\vec{\nabla} f)$. By definintion, the Laplacian of a scalar function f is equal to the divergence of the gradient of the function f, i.e.:

$$\text{div}\cdot(\overrightarrow{grad}\ f) = \vec{\nabla} \cdot (\vec{\nabla} f) = \nabla^2 f = \Delta f \qquad \text{[A2.37]}$$

It combines and links the static description of a field (described by its gradient) to the dynamic effects (the divergence) of this field in space and time.

It physically corresponds to a measure of the difference between the value of f at any point M and the mean value of f about point P. It can also be interpreted as the local mean curvature. The second derivative (or curvature) thus represents the local deviation of the mean with respect to the value at the point considered.

Like the nabla, divergence and rotational operators, the Laplacian is involved in many vector calculations in the field of electromagnetism (for example the Poisson and Laplace equations). The expressions of the Laplacian in the various coordinate systems are as follows.

A2.2.4.1. *Cartesian coordinates*

Scalar function $f(x, y, z)$; direct trihedron ($\vec{e}_x$, $\vec{e}_y$, $\vec{e}_z$):

$$\nabla^2 f = \frac{\partial^2 G_x}{\partial x^2} + \frac{\partial^2 G_y}{\partial y^2} + \frac{\partial^2 G_z}{\partial z^2} \qquad \text{[A2.38]}$$

A2.2.4.2. *Cylindrical coordinates*

Scalar function $f(\rho, \theta, z)$; direct trihedron ($\vec{e}_\rho$, $\vec{e}_\theta$, $\vec{e}_z$):

$$\nabla^2 f = \frac{1}{\rho}\frac{\partial}{\partial \rho}\left(\rho \frac{\partial f}{\partial \rho}\right) + \frac{1}{\rho^2}\frac{\partial^2 f}{\partial \theta^2} + \frac{\partial^2 f}{\partial z^2} \qquad \text{[A2.39]}$$

A2.2.4.3. *Spherical coordinates*

Scalar function $f(r, \theta, \varphi)$; direct trihedron ($\vec{e}_r$, $\vec{e}_\theta$, $\vec{e}_\varphi$):

$$\nabla^2 f = \frac{1}{r^2}\frac{\partial}{\partial r}\left(r^2 \frac{\partial f}{\partial r}\right) + \frac{1}{r^2 \sin\theta}\frac{\partial}{\partial \theta}\left(\sin\theta \frac{\partial f}{\partial \theta}\right) + \frac{1}{r^2 \sin^2\theta}\frac{\partial^2 f}{\partial \varphi^2} \qquad \text{[A2.40]}$$

A2.2.5. *Relations in vector algebra*

The vector fields $\vec{A}$, $\vec{B}$ and $\vec{C}$ satisfy the following properties:

$$\vec{A} \wedge \vec{B} = -\vec{B} \wedge \vec{A} \qquad \text{[A2.41]}$$

$$\vec{A} \cdot (\vec{B} \wedge \vec{C}) = \vec{B} \cdot (\vec{C} \wedge \vec{A}) = \vec{C} \cdot (\vec{A} \wedge \vec{B}) \qquad \text{[A2.42]}$$

$$\vec{A} \wedge \vec{B} \wedge \vec{C} = \vec{B} \cdot (\vec{A} \cdot \vec{C}) - \vec{C} \cdot (\vec{A} \cdot \vec{B}) \qquad \text{[A2.43]}$$

If f and g are any two scalar functions, then the following properties are also verified:

$$\vec{\nabla}(f+g)=\vec{\nabla}f+\vec{\nabla}g \qquad \text{[A2.44]}$$

$$\vec{\nabla}(fg)=f\vec{\nabla}g+g\vec{\nabla}f \qquad \text{[A2.45]}$$

$$\vec{\nabla}\cdot(f\vec{A})=\vec{\nabla}f\cdot\vec{A}+f\vec{\nabla}\cdot\vec{A} \qquad \text{[A2.46]}$$

$$\vec{\nabla}\cdot(\vec{A}\wedge\vec{B})=\vec{B}\cdot\vec{\nabla}\wedge\vec{A}-\vec{A}\cdot\vec{\nabla}\wedge\vec{B} \qquad \text{[A2.47]}$$

$$\vec{\nabla}\wedge(f\vec{A})=\vec{\nabla}f\wedge\vec{A}+f\vec{\nabla}\wedge\vec{A} \qquad \text{[A2.48]}$$

$$\vec{\nabla}\wedge(\vec{\nabla}\wedge\vec{A})=\vec{\nabla}(\vec{\nabla}\cdot\vec{A})-\vec{\nabla}^2\vec{A}=\vec{\nabla}(\vec{\nabla}\cdot\vec{A})-\Delta\vec{A} \qquad \text{[A2.49]}$$

$$\vec{\nabla}\cdot(\vec{\nabla}f)=\vec{\nabla}^2 f=\Delta f \qquad \text{[A2.50]}$$

$$\vec{\nabla}\cdot(\vec{\nabla}\wedge\vec{A})=0 \quad ; \quad \vec{\nabla}\wedge(\vec{\nabla}f)=0 \qquad \text{[A2.51]}$$

A2.3. Integral transform theorems

A2.3.1. *Stokes' theorem*

Stokes' theorem (also known as the rotational theorem) links the flux of a vector field $\vec{G}$ across a closed surface to the rotational of the field:

$$\oint_{(C)}\vec{G}\cdot\vec{dl}=\iint_S(\vec{\nabla}\wedge\vec{G})\cdot\vec{dS} \qquad \text{[A2.52]}$$

A2.3.2. *Ostrogradsky's theorem*

Ostrogradsky's theorem (also known as the divergence theorem) links the flux of a vector field $\vec{G}$ across a closed surface to the divergence of the field:

$$\oiint_S\vec{G}\cdot\vec{dS}=\oiiint_\tau(\vec{\nabla}\cdot\vec{G})\cdot d\tau \qquad \text{[A2.53]}$$

A2.4. Fundamental relations

The scalar Laplacian is the divergence of the field gradient:

$$div\left(\overrightarrow{grad}\right) = \Delta \qquad [A2.54]$$

The divergence of the rotational of a field is always zero:

$$div\left(\overrightarrow{rot}\right) = 0 \qquad [A2.55]$$

The rotational of a field rotational is equal to the gradient of the divergence of this field minus the Laplacian:

$$\overrightarrow{rot}\left(\overrightarrow{rot}\right) = \overrightarrow{grad}\left(div\right) - \Delta \qquad [A2.56]$$

The rotational of the field gradient is zero:

$$\overrightarrow{rot}\left(\overrightarrow{grad}\right) = \vec{0} \qquad [A2.57]$$

Appendix 3

Frequency Spectrum[1]

A3.1. Introduction

Spectrum management at the international level is the responsibility of the ITU-R (International Telecommunication Union – Radiocommunication Sector), headquartered in Geneva. It regularly publishes a frequency allocation table whose role is to share the spectrum according to its use (fixed, mobile, amateur, radio astronomy, etc.) and geographic location worldwide. It is integrated into the Radio Regulations (RR).

Here we provide an overview by frequency range, emphasizing for each, the atmospheric and terrestrial influence, the system considerations and the associated services).

A3.2. The different frequency ranges

A3.2.1. *ELF waves (frequency less than 3 kHz)*

– Wavelength (λ): above 100 km.

– Atmospheric influence: the ionosphere forms the upper limit of the propagation waveguide.

– Terrestrial influence: the surface of the Earth constitutes the lower limit of the propagation waveguide.

– System consideration:

- need for very long antennas (linked to the wavelength);

For a color version of all figures in this book, see www.iste.co.uk/favennec/electromagnetic2.zip.

1. Hall, M.P.M. (1996). Overview of radiowave propagation. In *Radiowave Propagation*, Hall, M.P.M., Barclay, L.W., Hewitt, M.T. (eds). Institution of Electrical Engineers, London.

- very slow information flow;

- sensitivity to noise: antenna and atmospheric (distant stormy discharges, local electrical discharges).

– Associated services:

- submarine communications (short and long distances);

- subterranean communications;

- ultra-sensitive communications on a global scale;

- internal remote sensing of the Earth's subsoil.

A3.2.2. *VLF waves (3–30 kHz)*

– Wavelength(λ): 10–100 km (myriametric waves).

– Atmospheric influence:

- the ionospheric D layer forms the upper limit of the propagation waveguide.

– Terrestrial influence:

- the surface of the Earth constitutes the lower limit of the propagation waveguide;

– System consideration;

- need to raise very long antennas (100 m is only a small fraction of the wavelength);

- difficulty in constructing directional antennas;

- very slow flow of information.

– Associated services:

- world maritime telegraphy;

- long-distance fixed services;

- navigation system (OMEGA);

- standard time diffusion;

- subterranean communications;

- submarine communications;

- geological surveillance.

A3.2.3. *LF waves (30–300 kHz)*

– Wavelength(λ): 1–10 km (kilometric waves).

– Atmospheric influence:

- sky wave in the propagation waveguide formed by the Earth (lower limit) and the ionospheric D layer (upper limit) for frequencies below 100 kHz;

- sky wave and ground wave at frequencies above 100 kHz;

– Terrestrial influence: ground wave is influenced by the Earth's curvature.

– System consideration:

- need to raise very long antennas (100 m is only a small fraction of the wavelength);

- difficulty in building directional antennas.

– Associated services:

- long distance maritime communications;

- fixed services over long distances;

- broadcasting;

- radio navigation.

A3.2.4. *MF waves (300–3,000 kHz)*

– Wavelength(λ): 100–1,000 m (hectometric waves).

– Atmospheric influence:

- ground wave for short distances and lowest frequencies;

- sky wave for long distances and the highest frequencies. The night field is higher even for short distances.

– Terrestrial influence:

- reflection of the wave on the ground.

– System consideration:

- possibility of using directional antennas (multiple element antennas, L or T inverted antennas);

– Associated services:

- broadcasting;

- radio navigation;

- mobile communications (terrestrial, maritime, aeronautical);

- fixed services.

A3.2.5. *HF waves (3–30 MHz)*

– Wavelength(λ): 10–100 m (decametric waves).

– Atmospheric influence:

- ground wave for short distances particularly above the sea;

- sky wave for long distances (beyond the first hop: 1E or 1F).

– Terrestrial influence:

- reflection and scattering.

– System consideration:

- use of log-periodic antenna network (vertical or horizontal);

- use of vertical whip antenna;

- use of horizontal dipole antenna.

– Associated services:

- point-to-point fixed services;

- mobile services (aeronautical, terrestrial, maritime);

- long distance broadcasting.

A3.2.6. *VHF waves (30–300 MHz)*

– Wavelength(λ): 1–10 m (metric waves).

– Atmospheric influence:

- refraction and reflection due to irregularities in the refractive index permitting trans-horizon links;

- ionospheric scattering (on the sporadic layer E in particular) allowing trans-horizon links;

- Faraday rotation and ionospheric scintillation on Earth-space links.

– Terrestrial influence:

- mask effects (hills, mountains);

- diffraction inside valleys;

- reflection on lakes, the sea and flat terrain causing multi-paths effects.

– System consideration:

- use of multi-element dipole antennas (yagi antenna);

- use of slot and helicoidal antenna.

– Associated services:

- broadcasting (up to 100 km);

- mobile communications (terrestrial, aeronautical, maritime);

- mobile telephone;

- aeronautical radio navigation beacon.

A3.2.7. *UHF waves (300–3,000 MHz)*

– Wavelength(λ): 10–100 cm (decimetric waves).

– Atmospheric influence:

- atmospheric refraction;

- reflection on the atmospheric layers at the lowest frequencies;

- ducting effect at higher frequencies;

- fluctuations in the refractive index;

- forescatter beyond the horizon at frequencies above 500 MHz.

– Terrestrial influence:

- mask effects (hills, buildings).

– System consideration:

- use of multi-element dipole antennas (yagi antenna);
- possible use of broadband systems;
- use of reflector antennas at the highest frequencies.

– Associated services:

- television;
- aeronautical navigation;
- radars;
- point to point fixed services;
- mobile communications;
- telemetry;
- cellular communications.

A3.2.8. *SHF waves (3–30 GHz)*

– Wavelength(λ): 1–10 cm (centimetric waves).

– Atmospheric influence:

- attenuation by hydrometeors (rain, snow, hail);
- refraction;
- duct propagation;
- scintillation due to fluctuations in the refractive index.

– Terrestrial influence:

- diffraction by buildings;
- mask effect (buildings, etc.);
- reflection and scattering by the environment (buildings, trees, etc.).

– System consideration:

- use of satellite dishes or horns;
- use of waveguide;
- possibility of using several channels on each carrier.

– Associated services:

- fixed services (voice, multi-channel television);

- radar;

- fixed Earth-satellite link;

- mobile services;

- satellite telemetry.

A3.2.9. *EH waves (30–300 GHz)*

– Wavelength(λ): 1–10 mm (millimetric waves).

– Atmospheric influence:

- significant attenuation by hydrometeors (rain, snow, hail);

- attenuation by clouds, mist, dust, smoke, etc.;

- scattering by hydrometeors (rain, snow, hail);

- scintillations due to fluctuations in the refractive index;

- absorption by oxygen and water vapor.

– Terrestrial influence:

- mask effect (buildings, trees).

– System consideration:

- use of small parabolic antennas;

– Associated services:

- short line-of-sight communications (fixed or mobile);

- satellite remote sensing.

A3.2.10. *Sub-EHF waves (300–3,000 GHz)*

– Wavelength(λ): 0.1–1 mm (sub-millimetric waves).

– Atmospheric influence:

- significant attenuation by meteors (rain, snow, hail, clouds, dust, smoke, etc.);

- appearance of gradient of the refractive index (mirage);

- scintillations due to fluctuations in the refractive index;

- absorption by atmospheric gases (oxygen, water vapor, etc.).

– Terrestrial influence:

- mask effect (trees).

– System consideration:

- use of reflector or lens antennas.

– Associated services:

- short range line-of-sight communications;

- remote sensing.

A3.2.11. *Infrared waves (3–430 THz) and light waves (430–860 THz)*

– Wavelength(λ): 0.1 mm–0.4 μm (infrared and light waves).

– Atmospheric influence:

- significant attenuation by meteors (rain, snow, hail, clouds, dust, smoke, etc.);

- appearance of gradient of the refractive index (mirage);

- scintillations due to fluctuations in the refractive index;

- absorption by atmospheric gases (carbon dioxin, etc.).

– Terrestrial influence:

- mask effect (small objects).

– System consideration:

- use of reflector or lens antennas;

- use of laser.

– Associated services:

- short path communications;

- indoor communications.

Appendix 4

The Decibel

A4.1. Introduction

The decibel (dB) is one tenth of a bel, named after Alexander Graham Bell, inventor of the telephone in 1876. It is a logarithmic unit of measure for the ratio between two powers, two voltages, two pressures, etc. It is used more particularly in the fields of telecommunications, electronics, acoustics, etc. It does not belong to the international system (SI) of units.

A4.2. Definition

It is defined in two different ways depending on whether we consider voltages (V) or powers (P).

In the case of voltages, the variable X in dB is defined by 20 times the logarithmic base 10 of the ratio of the output voltage (V_{out}) to the input voltage (V_{in}):

$$X_{dB} = 20 \log_{10} \left(V_{out} / V_{in} \right) \quad [A4.1]$$

In the case of power, the variable X in dB is defined by 10 times the logarithmic base 10 of the ratio of the output power (P_{out}) to the input power (P_{in}):

$$X_{dB} = 10 \log_{10} \left(P_{out} / P_{in} \right) \quad [A4.2]$$

We will use the term amplification or attenuation depending on whether the variable X_{dB} is positive or negative.

For a color version of all figures in this book, see www.iste.co.uk/favennec/electromagnetic2.zip.

Electromagnetic Waves 2,
coordinated by Pierre-Noël FAVENNEC.

A4.3. The different variants

Different variants exist (dB, dBm, dBW, dBi, dBc, dBµV, dBµVm^{-1}) depending on the references taken by the input voltages and the input powers:

– dB: the reference power is the input power:

$$X_{dB} = 10\log_{10}\left(P_{out} / P_{in}\right) \quad [A4.3]$$

– dBm: the reference power (the input power) is equal to one milliwatt (1 mW):

$$X_{dB} = 10\log_{10}\left(P_{out} / 1mW\right) \quad [A4.4]$$

– dBW: the reference power (the input power) is equal to one watt (1 W):

$$X_{dB} = 10\log_{10}\left(P_{out} / 1W\right) \quad [A4.5]$$

– dBi: here we refer to the gain of an antenna compared to an isotropic aerial. An isotropic aerial is an antenna that emits the same amount of energy in all directions. In reality, such an antenna does not exist:

$$X_{dB} = 10\log_{10}\left(P_{antenna} / P_{isotropicantenna}\right) \quad [A4.6]$$

– dBd: here we refer to the gain of an antenna compared to a half-wave dipole aerial. This aerial has a physical reality:

$$X_{dB} = 10\log_{10}\left(P_{antenna} / P_{dipoleantenna}\right) \quad [A4.7]$$

NOTE.– Catalogs do not generally specify whether they refer to antenna gains in dBi or in dBd. Yet the difference is important. A 10 dBd gain antenna has a gain of 12.15 dBi. We note that there is a great temptation to display gains in dBi rather than gains in dBd without specifying the unit used.

– dBc: the reference power (the input power) is equal to the power of the carrier (P_c) where "c" is the carrier:

$$X_{dB} = 10\log_{10}\left(P_{out} / P_c\right) \quad [A4.8]$$

– dBµV: the reference voltage (the input voltage) is equal to one µVolt (1 µV):

$$X_{dB} = 20\log_{10}\left(Voltage_{out} / 1\mu V\right) \quad [A4.9]$$

– dBµVm^{-1}: the reference field (the input field) is equal to 1 µVm^{-1}:

$$X_{dB} = 10\log_{10}\left(Field_{out} / 1\mu Vm^{-1}\right) \quad [A4.10]$$

A4.4. Decibel operations

As a reminder, we have the following relationships on logarithms:

$$\log_{10}(ab) = \log_{10}(a) + \log_{10}(b) \quad [A4.11]$$

$$\log_{10}(a/b) = \log_{10}(a) - \log_{10}(b) \quad [A4.12]$$

$$\log_{10}\left(a^b\right) = b\log_{10}(a) \quad [A4.13]$$

Also if:

$$A_{dB} = 10\log_{10}(P_{out} / P_{in}) = 10\log_{10}(R) \quad [A4.14]$$

then:

$$R = 10^{\frac{A}{10}} \quad [A4.15]$$

NOTE.– The advantage of using the dB scale is that the gains (amplifications) or attenuations are added up instead of multiplied.

A4.5. Correlation table

R	R (scientific)	A (dB)
0.00001	10^{-5}	– 5
0.0001	10^{-4}	– 4
0.001	10^{-3}	– 3
0.01	10^{-2}	– 2
0.1	10^{-1}	– 1

R	R (scientific)	A (dB)
1	1	0
10	10^{+1}	1
100	10^{+2}	2
1,000	10^{+3}	3
10,000	10^{+4}	4

A4.6. Particular values

Let us consider some particular ratio values, namely 0.1 (1/10), 0.5 (1/2), 2 and 10, giving:

$$10\log_{10}(1/10) = -10 \quad [A4.16]$$

$$10\log_{10}(1/2) = -3 \quad [A4.17]$$

$$10\log_{10}(2) = 3 \quad [A4.18]$$

$$10\log_{10}(10) = 10 \quad [A4.19]$$

Each time we divide by 10 in the current scale, we subtract 10 in the decibel scale.

Whenever we divide by 2 in the current scale, we subtract 3 in the decibel scale.

Appendix 5

The International Visibility Code

Weather condition	Precipitation		mm/h	Visibility (m)	Attenuation (db/km)
Dense fog				0	
				50	315
Thick fog				200	75
Moderate fog				500	28.9
Light fog	Snow	Thunderstorm	100	770	18.3
Very light fog	Snow			1,000	13.8
	Snow	Heavy rain	25	1,900	6.9
	Snow			2,000	6.6
Light mist	Snow	Moderate rain	12.5	2,800	4.6
	Snow			4,000	3.1
Very light mist	Snow	Light rain	2.5	5,900	2
				10,000	1.1
Clear weather		Drizzle	0.25	18,100	0.6
				20,000	0.54
Very clear weather				23,000	0.47
				50,000	0.19

For a color version of all figures in this book, see www.iste.co.uk/favennec/electromagnetic2.zip.

Electromagnetic Waves 2,
coordinated by Pierre-Noël FAVENNEC.

List of Acronyms and Constants

Acronyms

AR *Axial Ratio*

BER *Bit Error Rate*

CAD *Computer-Aided Design*

CEM *Compatibilité Électromagnétique (Electromagnetic Compatibility (EMC))*

CP *Circular Polarization*

CPW *Coplanar Waveguide*

DCS *Digital Cellular System*

DRA *Dielectric Resonant Antenna*

EIRP *Effective Isotropic Radiated Power*

EM *Electromagnetic*

EMC *Electromagnetic Compatibility*

ERP *Effective Radiated Power*

GPS *Global Positioning System*

GSM *Global System for Mobile Communications*

HPBW *Half-Power Beamwidth*

ISI *Inter-Symbol Interference*

Constants

Boltzmann constant $k = 1.3805 \times 10^{-23}\ joules/Kelvin$

Speed of light in vacuum $c = 2.998 \times 10^{8}$ m.s^{-1}

Vacuum impedance $\eta_0 = 377\Omega$

Vacuum permeability $\mu 0 = 4\pi \times 10^{-7}$ H/m = $1/\varepsilon_0.c^2$

Vacuum permittivity $\varepsilon 0 = 8.85419 \times 10^{-12}$ F/m $= \dfrac{1}{\mu_0.c^2}$

References

Al-Jabr, A. (2009). A General ADE-FDTD algorithm for the simulation of different dispersive materials. *Forum Progress in Electromagnetics Research Symposuim*, 23–27.

Ammann, M. (2007). Non-Trivial Materials in EM-FDTD. PhD thesis, University of Zurich.

Balanis, C.A. (2008). *Modern Antenna Handbook*. John Wiley & Sons, New York.

Berenger, J.P. (1994). A perfectly matched layer for free-space simulation in finite-difference computer codes. *EUREM Symposium*, Bordeaux.

Berenger, J.P. (1996). Three-dimensional perfectly matched layer for the absorption of electromagnetic waves. *Journal of Computational Physics*, 127, 363–379.

Booton, R.C. (1992). *Computational Methods for Electromagnetics and Microwaves*. John Wiley & Sons, New York.

Cardoso, J.R. (2017). *Electromagnetics through the Finite Element Method*. CRC Press, Boca Raton.

Cartini, R. (1993). *Panorama encyclopédique des sciences*. Belin, Paris.

Chari, M.V.K., Salon, S.J. (2000). *Numerical Methods in Electromagnetism*. Academic Press, Cambridge.

Chew, W.C., Jin, J., Lu, C., Michielssen, E., Song, J.N. (1997). Fast solution methods in electromagnetics. *IEEE Transactions Antennas Propagation*, 45, 533–543.

Chew, W.C., Jin, J.-M., Michielssen, E., Song, J.N. (2001). *Fast and Efficient Algorithms in Computational Electromagnetics*. Artech House, London.

Choi, S., Sarabandi, K. (2010). Design of efficient terahertz antenna: Carbon nanotube versus gold. *2010 IEEE Antennas and Propagation Society International Symposium*, Toronto.

Chu, L.J. (1948). Physical limitations of omni-directional antennas. *Journal of Applied Physics*, 19, 1163–1175.

Electromagnetic Waves 2,
coordinated by Pierre-Noël FAVENNEC.

Clarricoats, P.J.B., Olver, A.D. (1984). *Corrugated Horns for Microwave Antennas*. Peter Peregrinus, London.

Collin, R.E. (1991). *Field Theory of Guided Waves*. IEEE Press, New York.

Collin, R.E., Zucker, F.J. (1969). *Antenna Theory*. McGraw-Hill, New York.

Courant, R., Friedrichs, K., Lewy, H. (1967). On the partial difference equations of mathematical physics. *IBM Journal of Research and Development*, 11(2), 215–234.

Davidson, D.B. (2014). *Computational Electromagnetics for RF and Microwave Engineering*. Cambridge University Press, New York.

de Fornel, F., Favennec, P.-N. (eds) (2007). Mesures en électromagnétisme. *Revue RS série I2M*, 7(1–4).

Dettman, J.W. (1962). *Mathematical Methods in Physics and Engineering*. McGraw-Hill, New York.

Drabowitch, S., Ancona, C. (1978). *Antennes 2 : applications*. Masson, Paris.

Du Hamel, R., Isabell, D. (1957). Logarithmically periodic antenna designs. *National Convention Record*, Los Angeles.

Dussopt, L., Lamy, Y., Joblot, S., Lantéri, J., Salti, H., Bar, P., Sibuet, H., Reig, B., Carpentier, J.-F., Dehos, C., Vincent, P. (2012). Silicon interposer with integrated antenna array for millimeter-wave short-range communications. *2012 IEEE/MTT-S International Microwave Symposium Digest*, Montreal.

Eom, H.J. (2004). *Electromagnetic Wave Theory for Boundary Value Problems: An Advanced Course on Analytical Methods*. Springer, Berlin.

EM Software & Systems (2009). –S.A. (Pty) Ltd, Stellenbosch, South Africa, "FEKO –Field Computational Involving Bodies of Arbitrary Shape", Suite 5.5. Available at: http://www.feko.info.

Er-Ping, L. (2009) Computational Electromagnetics for Electromagnetic Compatibility/Signal Integrity Analysis. IEEE EMC DL, University of Missouri–St. Louis.

Ergûl, Ô., Gürel, L. (2006). On the accuracy of MFIE and CFIE in the solution of large electromagnetic scattering problems. *European Conference on Antennas and Propagation*, Nice.

Favennec, P.-N. (2008). *Mesures de l'exposition humaine aux champs radioélectriques – environnement radioélectrique*. Techniques de l'Ingénieur, Saint-Denis.

Felber, P. (2000). Fractal antennas, A literature study as a project for ECE 576. Illinois Institute of Technology [Online]. Available at: http://www.ece.iit.edu/~pfelber/fractalantennas.pdf.

Friis, H.T. (1944). Noise figures of radio receivers. *Proceedings of the IRE*, 32(7), July.

Garg, R. (2008). *Analytical and Computational Methods in Electromagnetics*. Artech House, Norwood.

Gerl, M., Janot, C. (1970). *Relativité électromagnétisme*. Hachette, Paris.

Geyi, W. (2015). *Foundations for Radiofrequency Engineering*. World Scientific Publishing, Singapore.

Ghannay, N., Romdhani, F., Samet, A. (2008). Caractérisation des structures par utilisation des fonctions de base RWG dans la méthode des moments. In *5ème Conférence international d'électronique et d'automatique*. Hammamet.

Gibson, W.C. (2008). *The Method of Moments in Electromagnetics*. Chapman & Hall, London.

Grange, F. (2010). Matériaux composites pour antenne miniature. PhD thesis, University of Rennes 1.

Hagdahl, S. (2005). Hybrid Methods for Computational Electromagnetics in Frequency Domain. BA thesis, Royal Institute of Technology, Stockholm.

Hao, Y., Mittra, R. (2009). *FDTD Modeling of Metamaterials, Theory and Applications*. Arthech House, Boston.

Harrington, R.F. (1993). *Field Computation by Moment Methods*. IEEE Press, Hoboken.

Harrington, R.F. (2001). *Time-harmonic Electromagnetic Fields*. McGraw-Hill, New York.

Hodges, R.E., Rahmat-Samii, Y. (1997). The evaluation of MFIE integrals with the use of vector triangle basis functions. *Microwave and Optical Technology Letters*, 14(1), 9–14.

Houssemeddine, B.E.M., Torres, F., Zakka El Nashef, G., Barelaud, B., Ngoya, E. (2013). Antenne intégrée sur silicium pour applications millimétriques System on Chip (SoC). *18èmes Journées Nationales Microondes*, 15–17 May, Paris [Online]. Available at: www.Top500.org.

Huang, Y., Boyle, K. (2008). *Antennas from Theory to Practice*. John Wiley & Sons, New York.

Hubing, T., Su, C., Zeng, H., Ke, H. (2008). Survey of current computational electromagnetics techniques and software. Technical report, Clemson University.

James, J.R., Hall, P.S. (1989). *Handbook of Microstrip Antennas*. Peter Peregriuns, London.

Jin, J. (2002). *The Finite Element Method in Electromagnetics*. John Wiley & Sons, New York.

Jin, J. (2014). *The Finite Element Method in Electromagnetics*, 3rd edition. John Wiley & Sons, New York.

Joannopoulos, J.D., Robert, D.M., Joshua, N.W. (1995). *Photonic Crystals.* Princeton University Press, Princeton.

Jung, B.H., Sarkar, T.K. (2002). Transient scattering from three dimensional conducting bodies by using magnetic field integral equation. *Journal of Electromagnetic Waves and Applications*, (16)1, 111–128.

Keller, J.B., Lewis, R.M., Seckler, B.D. (1956). Asymptotic solution for some diffaction problems. *Communications on Pure and Applied Mathematics*, 9, 207.

Kittel, C. (2007). *Physique de l'état solide.* Dunod, Paris.

Konrad, A., Silvester, P. (1973). Triangular finite elements for the generalized Bessel equation of order *m*. *The International Journal for Numerical Methods in Engineering*, 7, 43–55.

Kouyoumjian, R.G. (1974). A uniform geometrical theory of diffraction for an edge in perfectly conducting surface. *Proceedings of the IEEE*, 62(11), 1448.

Kraus, D., Fleisch, D.A. (1999). *Electromagnetic with Application.* McGraw-Hill, New York.

Lavrinenko, A.V., Lægsgaard, J., Gregersen, N., Schmidt, F., Søndergaard, T. (2015). *Numerical Methods in Photonics.* CRC Press, Boca Raton.

Liu, S., Yuan, N., Mo, J. (2003). A novel FDTD formulation for dispersive media. *IEEE Microwave and Wireless Components Letters*, 13(5), 187–189.

McDonald, B.H., Wexler, A. (1972). Finite-element solution of unbounded field problems. *IEEE Transactions on Microwave Theory and Techniques*, 20(12), 841–847.

Milligan, T.A. (2005). *Modern Antenna Design.* John Wiley & Sons, New York.

Mitin, V.M., Dmitry, I.S. (2017). *An Introduction to Applied Electromagnetics and Optics.* CRC Press, Boca Raton.

Moharos, I., Oldal, I., Szekrényes, A. (2012). *Finite Element Method.* University of Budapest, Budapest.

Molinet, F., Andronov, I., Bouche, D. (2005). *Asymptotic and Hybrid Methods in Electromagnetics.* The Institution of Engineering and Technology, London.

Mosig, J.R. (2007). Rayonnement et antennes, Report. Institut "Transmission Ondes et Photonique" (LEMA-ITOP-EPFL), Lausanne.

Pereda, J.A., Vegas, A., Prieto, A. (2002). FDTD modeling of wave propagation in dispersive media using the mobius transformation technique. *IEEE Transactions on Microwave Theory and Techniques*, 50(7), 1689–1695.

Peterson, A.F., Ray, S.L., Mittra, R. (2001). *Computational Methods for Electromagnetics.* IEEE Press, New York.

Picon, O. (2009). *Les antennes*. Dunod, Paris.

Pozar, D.M. (1997). *Microwave Engineering*. John Wiley & Sons, New York.

Pozar, D.M. (2009). New results for minimum Q, maximum gain, and polarization properties of electrically small arbitrary antennas. In *3rd European Conference on Antennas and Propagation*. Berlin.

Rao, S.M., Glisson, A.W., Wilton, D.R. (1982). Electromagnetic scattering by surfaces of arbitrary shape. *IEEE Transactions on Antennas and Propagation*, 409–418.

Rokhlin, V., Coifman, R.R., Wandzura, M.S. (1993). The fast multipole method: A pedestrian prescription. *IEEE Transactions on Antennas and Propagation*, 35, 7–12.

Rudge, A.W., Milne, K., Olver, A.D., Knight, P. (1982). *The Handbook of Antenna Design*. Peter Peregrinus, London.

Rumsey, V.H. (1957). *Frequency Independent Antennas*. Academic Press, Cambridge.

Safin, E. (2016). *Modal Analysis Methods for Small Antenna Design*. Kiel University, Kiel.

Saison-Francioso, O. (2014). Étude théorique et modélisation par la méthode FDTD de nanostructures plasmodiques : application à la conception de biocapteurs. PhD thesis, Institut d'électronique de micro- et de nanotechnologie, Lilles.

Sarhan, M.M. (2014). *Computational Nanotechnology Using Finite Difference Time Domain*. CRC Press, Boca Raton.

Sarychev, K.A., Shalaev, V.M. (2007). *Electrodynamics of Metamaterials*. World Scientific Publishing, Singapore.

Serres, M., Farouki, N. (1997). *Dictionnaire des sciences*. Flammarion, Paris.

Schneider, J.B. (2017). Understanding the finite-difference time-domain method [Online]. Available at: https://www.eecs.wsu.edu/~schneidj/ufdtd/ ufdtd.pdf.

Silver, S. (1984). *Microwave Antenna Theory and Design*. The Institution of Engineering and Technology, Stevenage.

Silvester, P. (1969). A general high-order finite-element waveguide analysis program. *IEEE Transactions on Microwave Theory and Techniques*, 17(4), 504–510.

Silvester, P. (1978). Construction of triangular finite element universal matrices. *International Journal of Numerical Methods in Engineering*, 12, 237–244.

Silvester, P., Ferrari, R.L. (1990). *Finite Elements for Electrical Engineers*. Cambridge University Press, Cambridge.

Sirenko, Y., Velychko, L. (2016). *Electromagnetic Waves in Complex Systems*. Springer, Berlin.

Skrievervik, A.K., Zürcher, J.F., Staub, O., Mosig, J.R. (2011). PCS antenna design: The challenge of miniaturization. *IEEE Antenna and Propagation*, 43(4).

Smith, D.G. (2013). *Field Guide to Physical*. SPIE Press, Bellingham.

SRSR (1994). means *Structures Rayonnantes à Symétrie de Révolution*, is a computational software for the analysis of radiating structures with rotational symmetry, developed by the Antennas Laboratory, Orange Labs, La Turbie, France, *Annales des Télécommunications*, 49(9–10), 575–588.

Stratton, J.A. (1941). *Electromagnetic Theory*. McGraw-Hill, New York.

Stutman, W.L. (1981). *Antenna Theory and Design*. John Wiley & Sons, New York.

Sullivan, D.M. (2000). *Electromagnetic Simulation Using the FDTD Method*. IEEE Press, New York.

Sumithra, P., Thiripurasundari, D. (2017). Review on computational electromagnetics. *Advanced Electromagnetics*, 6(1), 42–55.

Taflove, A. (1995). *Computational Electromagnetics*. Artech House, London.

Taflove, A. (2000). *Computational Electromagnetics*, 2nd edition. Artech House, London.

Taflove, A. (2005). *Computational Electromagnetics*, 3rd edition. Artech House, London.

Taflove, A., Brodwin, M.E. (1975). Numerical solution of steady state electromagnetic scattering problems using the time dependent Maxwell's equations. *IEEE*, 23(8).

Tapan, K.S. (2000). *Method of Moments Applied to Antennas*. Syracuse University, Syracuse.

Tarricone, L., Esposito, A. (2004). *Grid Computing for Electromagnetics*. Artech House, London.

Thourel, L. (1956). *Les antennes*. Dunod, Paris.

Tretyakov, S. (2003). *Analytical Modeling in Applied Electromagnetics*. Artech House, London.

Ufimtsev, P.Y. (1962). Method of edge waves in the physical theory of diffraction, Report. Defense Technical Information Center.

Ufimtsev, P.Y. (2014). *Fundamentals of the Physical Theory of Diffraction*. John Wiley & Sons, New York.

Valamparambil, S., Chew, W.C., Song, J. (2003). 10 million unknowns: Is that big? *IEEE Antennas Propagation*, (45)2, 43–58.

Vial, A., Grimault, S.A., Dimetrio, M., Barchiesi, D., Lamy de la Chapelle, M. (2005). Improved analytical fit of gold dispersion: Application to the modeling of extinction spectre with FDTD method. *Physical Review*, 71.

Volakis, J.L., Chatterjee, A., Kempel, L.C. (2001). *Finite Element Method for Electromagnetics Antennas, Microwave Circuits and Scattering Applications*. IEEE Press, New York.

Weiland, T. (1977). A discretization method for the solution of Maxwell's equations for six component fields. *Electronics and Communication*, 31, 116.

Wheeler, H. (1947). Fundamental limitations of small antennas. *Proceedings of the IEEE*, 1479–1484.

Yee, K.S. (1966). Numerical solution of initial boundary value problems involving Maxwell's equations in isotropic media. *IEEE Transaction in Antenna and Propagation*, 14(3), 302–307.

List of Authors

Jean-Pierre BLOT
EUROSATCOM
Les Ulis
France

Pierre-Noël FAVENNEC
ArmorScience
Lannion
France

Ibrahima SAKHO
University of Thiès
Senegal

Hervé SIZUN
ArmorScience
Lannion
France

Index

A, C

D, E

F, I

L, M

O, P

S, W

www.ingramcontent.com/pod-product-compliance
Lightning Source LLC
LaVergne TN
LVHW011754200926
848484LV00003B/707
9781789450071